LICENSED ARCHITECT

BEN G. WONG
No. C-016673
RENEWAL DATE

STATE OF CALIFORNIA

STRUCTURAL DESIGN
FOR ARCHITECTS

Structural Design for Architects

Alec Nash

NICHOLS PUBLISHING COMPANY : NEW YORK

of America in 1990 by

Nichols Publishing Company,
Post Office Box 96,
New York, N.Y. 10024

Library of Congress Cataloging-in-Publication Data
Nash, Alec
    Structural design for architects / Alec Nash.
        p.   cm.
    Includes bibliographical references.
    1. Structural design.   I. Title
    TH845.N37   1990
    824.1'771—dc20   89-77280
                        CIP

ISBN: 0-89397-366-1

Printed in Great Britain

Contents

Figures

Plates

Introduction

An architect is not usually responsible for producing detailed structural calculations and drawings, unless the building concerned is very small and simple. Where the architect can be most effective in the field of structural design is in the clarity of the manner in which suggested solutions, in the form of schematic designs, are put to a structural engineer.

It is vital that an architect can propose forms from which the structural engineer need not deviate, to the extent that the original design concept is violated. It is also important that he or she is able to make an informed and rational choice between apparently unrelated structural systems.

The theme of this book therefore arises from the necessity for an architect to possess an extensive structural vocabulary, based on a clear understanding of the relevant underlying principles. Although written mainly for practising architects, it is hoped that the book will also provide a fresh perspective on the subject for building surveyors as well as for civil and structural engineers.

The first two chapters cover the principles of statics and the nature of available structural materials in sufficient depth to permit a clear understanding of the following six chapters on structural form. Structural elements of increasing complexity are introduced, using examples from historical and modern works of architecture, as well as from nature.

References to codes of practice have been kept to a minimum, so that the book should have an international appeal. Where design stresses are essential to the understanding of the text, these have been included in the simplest possible form.

Chapter 1

PHYSICS AND MATHEMATICS IN ARCHITECTURE

A treatise which sets out to explore the role of structural design in architecture can only evolve out of a clear idea of the meaning of these two terms. It is a widely held belief in those societies which encourage early specialization in schools that architecture belongs in the domain of the arts whilst engineering, whether the discipline be concerned with machines, electricity, aviation or building structures, possesses a collective identity within the world of technology.

Such a dichotomy may work very well in the case of a surrealist painter at one extreme or the engineer responsible for a hidden production process at the other. There are, however, creations of the human intelligence which cannot be judged solely against the single criterion of function or of aesthetics. The practice and profession of architecture, by its very nature, is arguably the most striking example of a field of endeavour where an understanding rooted in both the arts and the sciences is essential.

If architecture is concerned with the spatial organization of man's activities, then the spaces that emerge from such organization will have their own physical or implied boundaries. In buildings, we call these physical boundaries walls, floors and roofs, often referred to as the enclosing elements of a building. It is not unreasonable to expect these enclosing elements to remain in position once the building is completed. The idea of structural design evolves from this hope of permanence.

Yet many structural elements whose explicit purpose is to maintain the physical integrity of the enclosed space are outstandingly beautiful. The column in Classical architecture and the vaulted ceiling in Gothic architecture are two particular forms which will be explored in later chapters. Neither of these forms can be dismissed simply as responses to the problems imposed by the physical constraints of existing within the earth's gravitational field. Ably as they fulfil this task, their forms arise from ideas more diverse than those engendered by structural demands alone. Gravity is, nevertheless, an inescapable condition of terrestrial existence, and provides a suitable starting point for the introduction of some of the laws of physics as they affect architecture.

The relevance of Newton's Laws to built forms

The structures conceived during the Gothic era in Northern European cathedral architecture occupied several centuries before the life span of Sir Isaac Newton (1642–1727). Yet the need to create equilibrium within a complex arrangement of forces was clearly understood, at least at an intuitive if not at a mathematical level. It is idle to speculate on whether the master masons responsible for deciding on the proportions of the structural elements thought of concepts such as force, gravity, and equilibrium in an abstract as well as a physical sense. Whatever the answer to that question, their judgements were in the majority of cases correct. The precise formulation of the laws of mechanics as they affect matter within a gravitational field had to wait for Newton's Three Laws.

Building on the earlier investigations of Johannes Kepler (1571–1630) and Galileo Galilei (1564–1642), Newton's Laws defined the relationship that exists between force, mass and acceleration. It is the idea of force that is of most importance in determining the nature and the proportions of structural members. The nature of force, however, can only be fully understood as the description of the experience of a mass trying to accelerate. Newton's Laws will first be stated, and then illustrated with reference to some familiar objects.

1.   Any body remains in a state of rest or uniform motion when no unbalanced force acts upon it.
2.   For a mass to undergo an acceleration, a force is required that is equal to the product of mass and acceleration.
3.   Every action must have an equal and opposite reaction.

The first law is of more relevance to the study of moving objects, classified by physicists as dynamics and kinematics. These enter the field of structural design when significant movements and oscillations have to be accommodated, such as those arising from wind forces or earthquakes.

In the second law, the term mass may be taken to mean a quantity of matter. The unit of mass is the kilogram. There will be as much mass in a kilogram of lead whether it is located on earth, on the moon, or in outer space. What will vary is the extent to which the lead will be constrained to move to another position. A kilogram of lead in outer space, being remote from any other objects, tends to be almost at rest, being under the influence of minimal unbalanced force. Its condition is very close to that described in Newton's First Law.

The reason for describing this condition as being almost, rather than totally, at rest lies in Newton's Law of Universal Gravitation, in which he proved that the force *(F)* by which one mass *(m₁)* attracts another mass *(m₂)* is proportional to the product of the masses, and inversely proportional to the square of the distance between them *(r)*. This can be stated algebraically as:

$$F = G \, m_1 \, m_2 \, / \, r^2$$

where *G* is the gravitational constant applying everywhere in the universe. Even in deepest space, therefore, there are other attractive masses, but the squares of their vast distances from the kilogram of lead will create forces of a very minute order.

Near to the moon, and to a greater extent close to the surface of the earth, the kilogram of lead will experience an appreciable force, and will accelerate according to the relationship expressed in Newton's Second Law. Although the earth is not a perfect sphere, the acceleration of any mass towards the centre of the earth *(g)* is very nearly constant for any point on the earth's surface, and has been confirmed by experiment to be an increase in velocity of 9.81 metres per second for every second, expressed mathematically as:

$$g = 9.81 \text{ m/s}^2$$

This property possessed by all objects of falling at the same rate had been established earlier by Galileo in his experiments conducted from the top of the Leaning Tower of Pisa. The implication in Newton's Second Law is that if the acceleration is constant, the force must vary as the mass. The unit of force adopted in the Système Internationale notation has been appropriately named the Newton, and is defined as that force which will cause a mass of one kilogram to accelerate by one metre per second per second. The consequence for objects within the earth's gravitational field can be expressed as:

1 Newton will accelerate 1 kilogram by   1 m/s²
∴ 9.81 Newtons will accelerate 1 kilogram by 9.81 m/s²

A mass of one kilogram, therefore, will exert a force of 9.81 Newtons towards the centre of the earth, that is downwards. In other words, one kilogram weighs about ten Newtons, which is a good enough approximation in structural design.

Loading on structures

The principle inherent in Newton's Third Law is that if objects are to have their accelerations prevented and thus remain at rest, the forces which they exert must be balanced by an equal force in the opposite direction. If their gravitational forces are the actions, then the upward forces provided by whatever supports those objects are the reactions. This is the condition of zero unbalanced force for the state of rest in Newton's First Law.

Forces imposed on all forms of structure must, if stability is to be achieved, eventually be balanced at ground level. A condition must also be reached whereby all of the intervening structural elements transferring those forces to the ground must themselves obey Newton's First and Third Laws. The first, and often the most tedious procedure in the day-to-day design of structures is the quantification of the forces which those structures have to resist. The resulting data is known as the loading, and is, by definition, expressed in units of force. Since the Newton is a relatively small quantity of force – it is the weight of about one-tenth part of a kilogram bag of sugar – it is easier to handle groups of one thousand Newtons at a time. Thus structural loads are always expressed in units of kilonewtons, i.e.

$$1 \text{ kilonewton} = 1000 \text{ Newtons}$$

To express the weight of one kilogram in kilonewtons, therefore,

$$1 \text{ kilogram weighs } 10 \text{ Newtons}$$
$$\therefore 1 \text{ kilogram weighs } 10/1000 \text{ kilonewtons}$$
$$= 1/100 \quad \text{kilonewtons}$$
$$= 1/10^2 \quad \text{kilonewtons}$$

which can be abbreviated to $1/10^2$ kN, or 10^{-2} kN

Design loads on structures are sub-divided into those loads which are present all of the time, and those which are present for only part of the time. The former are known as dead loads, being the weight of the materials which constitute the building. The latter are known as live loads, being the weights of the people and objects for whom and for which the building is to be provided. The snow load on the roof is also referred to as a live load. A third category is the force imposed by the wind.

Dead load

Dead loads can be evaluated fairly accurately from the known densities of the materials used. The density of a material is a measure of the quantity of matter in a given volume, i.e. mass per unit volume. It is expressed in units of kilograms per cubic metre. Since the design of structures is concerned about resistance to forces, it is more useful to know the equivalent weight of a unit volume of a material in the earth's gravitational field. This 'weight density' can be expressed in kilonewtons per cubic metre, and can be derived from a known mass density as follows:

$$\text{mass density of water} = 1000 \text{ kg/m}^3$$
$$\therefore \text{ weight density of water} = 1000/10^2 \text{ kN/m}^3$$
$$= 10 \text{ kN/m}^3$$

The numerical value of the weight density of a material is therefore one hundredth of that of its mass density.

Live load

In view of the transient nature of live loads, an estimate is needed as to what is the likely maximum loading that an element of a structure could be required to resist. This can only be acquired by a distillation of observation and experience, the results of which, after extensive statistical analysis, are contained in the relevant codes of practice. These live loads, or superimposed loads as they are sometimes called, reflect the greatest number of people likely to occupy a space designated for a particular use. In the case of a roof, the live load is a measure of the greatest depth of snow that could, given the most severe weather conditions, lay on its surface.

Examples of live loads currently stipulated in the current British Standard Code of Practice are:

Roof		0.75 kN/m²
Domestic floor		1.50 kN/m²
Classroom floor		3.00 kN/m²
Assembly hall	4.00 to	5.00 kN/m²
Factory floor	7.50 to	10.00 kN/m²

These loads are seen to be expressed as forces acting on a unit area of a horizontal surface. Clearly, this is an idealized condition that will rarely, if ever, exist in reality. This method is, however, convenient and its application does, when applied with informed engineering judgement, result in safe structures. There are certain structural arrangements in buildings in which the consequences of one or more spans being temporarily without its live load need to be investigated. One such case is the cantilevered beam, which will be examined in Chapter 7.

Wind loads

The effect of wind is expressed meteorologically in terms of its speed. This is a direct effect of movements of volumes of air with varying temperatures, but has a further variation at any given location with height above the ground. The greater this height, the greater the wind speed, because of the decreasing effect of friction with the ground. For the purposes of the design of structures, these wind speeds are converted into equivalent static loads. These, as with live loads, are expressed as forces per unit area. The most important effect of wind loads is clearly as a horizontal force applied to a vertical surface, but it is equally important to investigate the effect of the wind on a roof as the air travels across its surface. As the speed of the wind increases, the surrounding air pressure decreases. Since the air inside the building is at atmospheric pressure, there will be an unbalanced upward pressure. Since pressure is the force per unit area, there will exist an unbalanced upward force. Unless the dead load of the roof provides a downward force great enough to counter this imbalance, the roof structure will begin to accelerate upwards. This phenomenon is a major cause of structural failure where roofs have not been adequately anchored down into the walls. The combined effect of the roof's dead load and the anchorage detail should be sufficient to provide an adequate factor of safety against uplift. The occupant of the top storey may be alarmed to see the roof hovering above his head on a windy day, even though it may be a perfect manifestation of Newton's First Law.

The need for some mathematical awareness

> If you want to learn about nature, to appreciate nature, it is necessary to learn the language that she speaks in. (Professor Richard Feynman, 1965, p.58)

Feynman, in writing about the relationship between mathematics and physics, stated a truth which may be unpalatable to those who are not mathematically inclined. Fortunately, however, the level of mathematics required for the understanding of that branch of physics involving the majority of structural forms at rest is relatively simple. The principles of statics can be clearly defined by means of elementary trigonometry, linear and quadratic equations, and a familiarity with indices and graphs. These mathematical techniques will be explained in this chapter, and cross-referenced to relevant topics in later chapters.

Calculus will not be covered, but it will be necessary at some stages to describe the relevance of the slope of a graph. This will be covered in the text in the appropriate place.

Even if an architect does not intend to carry out any design calculations for himself, and relies totally on his consulting structural engineer, certain basic interdependencies can be expressed more succinctly in a mathematical rather than a verbal sense. For example, Newton's Law of Universal Gravitation needed 28 words, but only seven mathematical terms and symbols.

The concept of force has already been explained, so that it is appropriate at this stage to introduce those mathematical topics which are associated with it.

Forces

Vector and scalar quantities

A vertical force acts towards or away from the centre of the earth's mass as a consequence of the gravitational acceleration g, or 9.81 m/s^2. Forces do not always have to be vertical, as exemplified by the loads on a structure caused by the wind. A force can act in any direction, and therefore its mathematical description must be very precise about the direction in which it acts, as well as about its magnitude. Such quantities are known as vector quantities. It is meaningless to refer to a force of 50 kilonewtons without knowing where it is pointing. Vector quantities are in fact depicted as lines with arrows pointing somewhere. The description of the vertical force experienced by a mass on the earth's surface by the word 'weight' carries the image of the downward pointing arrow. People and objects are not weighed sideways.

Mass, on the other hand, is independent of direction and is known as a scalar quantity. A kilogram of apples is a sufficient description of one's requirements in a greengrocer's shop. Information about their intended direction when purchased is irrelevant. It is the acceleration vector applied to the scalar quantity of mass which creates the vector quantity of force. As Newton discovered for himself, an apple will experience the same gravitational acceleration as any other object. Unbeknown to Newton at the time, his name would, more than two hundred years after his death, become synonymous with the downward force vector associated with a small apple of mass $1/9.81$ kilograms, i.e.

1 Newton will accelerate 1 kilogram by 1m/s^2

∴ 1 Newton will accelerate $1/9.81$ kilogram by 9.81 m/s^2

Equilibrium

For Newton's Third Law to apply, the forces acting anywhere in a structure must balance. This may appear obvious where the load from a column resting on its foundation demands a vertical reaction of equal magnitude from the ground. It is less obvious in the middle of the span of a beam. In the latter case, the internal forces in the beam come into play. Being internal does not invalidate their status as forces.

Adopting the symbol V for vertical forces and H for horizontal forces, the consequences of Newton's Third Law for a system at rest can be expressed by the equations:

$$\Sigma V = 0$$
$$\Sigma H = 0$$

where the Greek capital letter Σ means 'sum of'. The first equation means that downward forces must be balanced by upward forces. The second means that forces acting in a direction from left to right must be balanced by forces acting from right to left. Where a force acts at an angle to the vertical and horizontal axes, it can be expressed as the vertical and horizontal components of that force.

Components of a force

Figure 1.1(a) shows a force vector F acting at an angle θ to the horizontal axis. Such a vector combines the force components acting upwards and to the right, known respectively as the vertical and horizontal components. The magnitude of these components can be evaluated by using the trigonometric ratios 'sine' and 'cosine' as illustrated in Figures 1.1(b) and 1.1(c).

In the right-angled triangle in 1.1(b), the sine of the angle θ is defined as the ratio of the length of the side opposite the angle θ to the length of the hypoteneuse. The cosine of θ is the ratio of the length of the side adjacent to θ to the length of the hypoteneuse. More briefly,

$$\sin \theta = \text{opposite/hypoteneuse, i.e. BC/AB}$$
$$\cos \theta = \text{adjacent/hypoteneuse, i.e. AC/AB}$$

As the hypoteneuse AB approaches the vertical position, the angle θ approaches $90°$, the length of BC will approach that of AB, $\sin \theta$ will approach 1, and $\cos \theta$ will approach 0. Similarly, as AB becomes nearly horizontal, θ will approach 0, making AC nearly as long as AB. At this limit, $\sin \theta$ approaches 0, and $\cos \theta$ approaches 1.

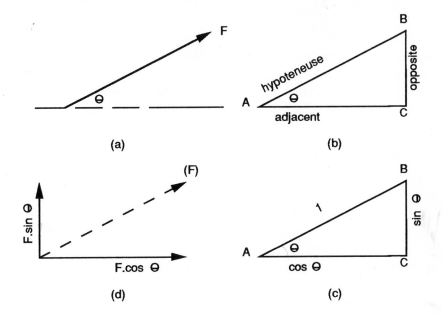

Figure 1.1 *Force components*

In Figure 1.1(c), the hypoteneuse AB has unit length
∴ since BC/AB = sin θ, BC = sin θ
and since AC/AB = cos θ, AC = cos θ

Figure 1.1(a) can now be modified as in Figure 1.1(d), in which the vertical and horizontal components of F are expressed as F. sin θ and F. cos θ respectively. F is known as the resultant of these two force components, because if these two components are applied simultaneously, the result is a force of magnitude F at an angle of θ to the horizontal. The three forces do not, of course, act at the same time. The action is mathematically described in terms of the resultant *or* the two components.

The triangle of forces

Yet Figures 1.1(a) and 1.1(d) do not depict conditions of equilibrium. They are merely different ways of expressing the precise nature of a particular action. If the equality of action and reaction is to be expressed in the same diagram, the arrangement of vectors would be as shown in Figure 1.2(a), in which the direction of the resultant force F is reversed. Here, it is called the equilibriant, because a state of equilibrium exists between it and the two components F.sin θ and F.cos θ. If these three vectors are shown with the end of each one joining onto the beginning of the next, they will triangulate in one of the forms shown in Figures 1.2(b) or 1.2(c). The only difference between these triangles is that in the first one, the vector arrows follow each other round in a clockwise direction, whilst in the second the direction is anti-clockwise. This does not matter, since both represent the state of equilibrium. The critical factor is that the three arrows in each set follow each other round in the same sense rather than in opposition. These two configurations are known as triangles of forces. They will appear again in profusion in Chapters 4 and 6.

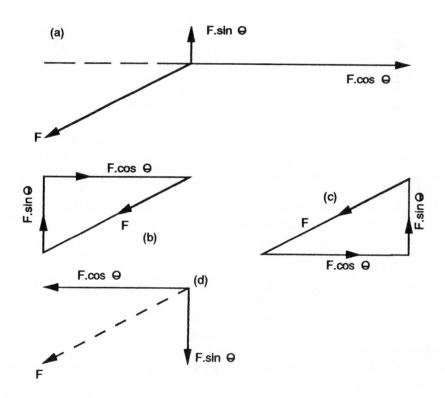

Figure 1.2   *Force triangulation*

The equilibrium conditions $\Sigma V = 0$ and $\Sigma H = 0$ can be seen to apply here if the equilibriant force F is resolved into its vertical and horizontal components. In Figure 1.2(d), the vector arrows of these two components are pointing in opposite directions to the two corresponding arrows of the components appearing in the triangle of forces.

The triangle of forces is the simplest form of a series of force diagrams known as polygons of forces. The number of sides in the polygon will be the same as the number of force vectors. If the vector sum of the vertical and horizontal components of all the forces each add up to zero, a closed polygon can be drawn, just as a closed triangle was drawn for the three forces in equilibrium. Such force diagrams will be explained at relevant points in subsequent chapters.

Bending moments

The moment of a force

It is an almost universal experience when opening a door to instinctively apply the force of one's hand at a position on the door most distant from the hinge. As with most intuitive or acquired ways of manipulating the physical world, there is an underlying logic rooted in the laws of physics. Given that a certain turning effect is required to open the door, it is intuitively felt that the nearer to the hinge the hand is applied, the greater will be the effort, or force, required. This turning effect of a force is called a moment.

A moment of a force is defined as the product of that force and the distance perpendicular to its line of action. The concept of the line of action of a force is particularly important, and is illustrated in Figures 1.3(a) and 1.3(b). In 1.3(a) the moment exerted by the force F could hardly be imagined to be anything other than the product of F and the distance x from the point P. In 1.3(b), the point of application of the force F is not on the same horizontal axis as the point P. If, however, the line representing the vector F is extended so that the distance x is perpendicular to it, the turning effect, or moment, of F about P will still be the product of F and x.

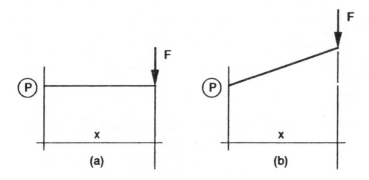

Figure 1.3 *Moment of a force*

The example of evaluating a force required to open a door is a problem of dynamics, involving the frictional and damping characteristics of the hinges. In architecture, for the most part, we are concerned with systems at rest, or statics. If the door were fixed to its frame on the hinge side in such a way that rotation in a horizontal plane were impossible, the moment would be transmitted to the door frame. The mechanism of a door would be converted to a simple cantilever, a structural form discussed in detail in Chapter 7. As a general principle, however, a structure is that which will prohibit movement rather than permit it. Forces and moments are absorbed into the structural system, and are major determinants of the nature and of the proportions of that system. This is true of structures in architecture or for any other purpose. This is not to say that structural elements do not undergo some dimensional changes in accepting those forces and moments. All structures deform under load, even though such deformations may not be perceptible without the use of instruments. But the salient condition in the design of structures is the state of equilibrium after these small deformations have taken place. The required sizes of a wide range of structural elements are determined solely from the necessity of absorbing, or resisting, a moment at a particular point.

The principle of moments

It was stated earlier in this chapter that in a system at rest, the equilibrium conditions $\Sigma V = 0$ and $\Sigma H = 0$ must apply. These two equations are sufficient to describe a state of equilibrium when all the forces involved meet at a point. Where, as in the case of the beam shown in Figure 1.4, the forces are not in the same vertical line, the purpose and function of the structural element is to move the downward load sideways to the positions where it is to be supported. Here, a third equation is necessary in which the net turning force on the system is also zero. This condition is expressed as $\Sigma M = 0$ where M is the algebraic symbol for moment. If the moments acting on the system in an anti-clockwise sense did not equal those acting in a clockwise sense, the system would be rotating. Instead of being a structure in equilibrium, the system would be a mechanism.

The magnitude of the load, or gravitational force, is given the algebraic symbol W. This type of load on a structure, representing a force concentrated at a particular point, is always referred to as a point load. In practice, such loads usually arise from a column unsupported from below, or from another beam spanning at right angles to this one. The load would always be spread over the relatively small width of this column or supported beam, but the convention of the vector arrow at one point is a good enough approximation.

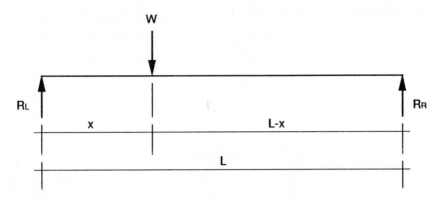

Figure 1.4   *Point load on beam*

Returning to Figure 1.4, the system of forces comprises the load W and the two reactions to this load at the ends of the beam. These reactions are given the symbols R_L and R_R, where

R denotes reaction
Subscript $_L$ locates the reaction at the left-hand support
Subscript $_R$ locates the reaction at the right-hand support

The horizontal dimension L is the span of the beam, and x is the horizontal distance between the lines of action of the vertical forces R_L and W. The corresponding distance between W and R_R is therefore $(L - x)$.

The magnitude of each of the reactions can be evaluated by using the two equilibrium conditions $\Sigma\,V = 0$ and $\Sigma\,M = 0$.

The third condition ($\Sigma\,H = 0$) is not required in this case. For vertical equilibrium, the three vertical forces must have an algebraic sum of zero, i.e.

$$W - R_L - R_R = 0$$

or, more conveniently expressed,

$$W = R_L + R_R$$
$$\text{(Downwards)} \quad \text{(Upwards)}$$

To describe the system as being in rotational equilibrium, i.e. where $\Sigma\,M = 0$, one of the support positions can be considered as a fulcrum, or visualized as analogous to the door hinge, while the two forces away from this position each exert a moment about that fulcrum. If the fulcrum is the left-hand support, the moments of forces R_R and W about that position must, for equilibrium, reflect the condition $\Sigma M = 0$. This process can be referred to as 'taking moments about R_L'. The force R_L will not exert a moment about this position because they coincide, so that there is no perpendicular distance between them. The equilibrium state can therefore be expressed as:

$$R_R \times L = W \times x$$
(Anti-clockwise)   (Clockwise)

In equations of this kind, the multiplication sign can be replaced by a dot, making the expressions easier to read where there are a large number of terms involved. This convention will be used from now on. The equation can thus be written as:

$$R_R . L = W . x$$

Dividing each side of the equation by L,

$$R_R = W . x / L$$

The magnitude of one of the reactions is therefore expressed in terms of the point load, its distance from the other support, and the span of the beam. This is an application of the principle of moments, which can be stated as: 'In any system at rest, the sum of the clockwise moments equals the sum of the anti-clockwise moments'.

If moments were taken about the right-hand support, the magnitude of R_L could be worked out in a similar manner:

$$R_L . L = W . (L - x)$$
(Clockwise)   (Anti-clockwise)
$$\therefore R_L = W . (L - x) / L$$

In explaining this principle, the units of the force and distance vectors have been omitted for clarity. In a problem involving actual forces and distances, the units should always be stated, if only to ensure that both sides of the equation are compatible. In problems of structural analysis encountered in architecture, applied loads are always in the S.I. system of units, expressed in kilonewtons. Spans of beams and dimensions relating to load positions, as well as heights of columns, are expressed in metres.

Simply supported beam with point load

In Figure 1.5(a), a load of ten kilonewtons is to be supported five metres from the left-hand support of a beam spanning fifteen metres. The beam is described as simply supported because the ends of the beam are assumed to be simply resting on their supports. No attempt is made to prevent the ends of the beam from rotating with respect to their supports as it takes up the load. This assumption is close to the reality of architectural details such as would occur at the junction of a steel beam and a vertical support constructed of a different material. There is no organic continuity between the horizontal and vertical structural elements.

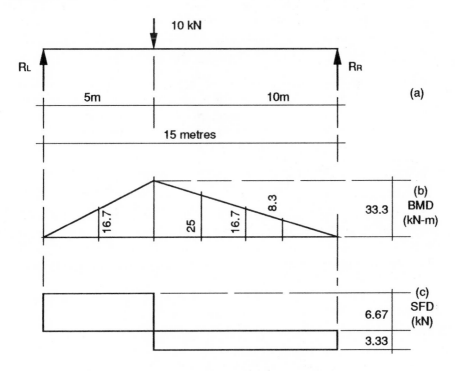

Figure 1.5   *Bending moment and shear force diagrams*

The self weight of the beam is ignored, so that there is a certain quality of abstraction in this problem. It is often necessary, however, when exploring alternative arrangements of structural elements, to examine the consequences of leaving out a column in the storey underneath the beam. By considering the load from the column in isolation, its influence on the dimensions of the beam can often be rapidly estimated.

By substituting the load and the salient dimensions into the values of the two reactions derived in the previous section, where

$$W = 10 \text{ kN}$$
$$L = 15 \text{ metres (15 m) and}$$
$$x = 5 \text{ metres (5 m) and}$$
$$(L - x) = (15 - 5) = 10 \text{ metres (10 m)},$$

$$R_R = 10 \times 5/15 = 3.33 \text{ kN and}$$
$$R_L = 10 \times 10/15 = 6.67 \text{ kN}$$

This calculation for the magnitude of the reactions can be checked by invoking the equilibrium condition for vertical forces, $\Sigma V = 0$.

As with all problems involving loads on beams, this means that the total downward load must equal the sum of the upward reactions. For this particular beam, the sum of the reactions is 10 kilonewtons, which is also the magnitude of the point load W. Unless two compensatory errors have been made in the calculations, the values of the reactions are correct.

It is at this stage that a distinction needs to be drawn between the external and internal equilibrium of a system. For the beam in Figure 1.5(a), the moments of forces causing clockwise rotation about a certain point will equal the moments of forces causing anti-clockwise rotation about that point. This is a necessary condition of rotational equilibrium, to which a beam in a structure must clearly subscribe. The point about which moments are taken can be anywhere along the beam, in keeping with the equilibrium condition $\Sigma M = 0$. If this condition were checked, for example, at the midspan position of the beam, that is 7.5 metres from either support, it would be expressed as

$$7.5.R_R = 7.5.R_L - (10 \times 2.5)$$
$$\text{(Anti-clockwise)}\quad\text{(Clockwise)}\quad\text{(Anti-clockwise)}$$
$$\therefore(3.33 \times 7.5) = (6.66 \times 7.5) - (10 \times 2.5)$$
$$\therefore\quad 25.0 = 50.0 - 25.0$$

which is arithmetically correct. The left-hand side of this equation expresses the net anti-clockwise moment, and the right-hand side the net clockwise moment. The minus sign before the (10×2.5) quantity appears because it is acting in an anti-clockwise sense as opposed to the clockwise sense of the (6.66×7.5) moment. They are both vector quantities, and must therefore be added algebraically.

The condition just described is that of external equilibrium. If, however, just one of the quantities separated by the equal sign were considered, there would be an imbalance. By covering up the right-hand half of the beam, the clockwise moment caused by the reaction at the left-hand support (R_L) and the load (W) remains as an unbalanced vector quantity, which can only be resisted by the material within the beam itself. The relationship between the unbalanced external moment and the internal moment within the beam will be pursued in Chapter 3, since it forms the basis of the design of all structural elements subjected to bending. For the present, it is enough to understand that this unbalanced external moment applied to the midspan position of the beam from the left-hand side is called the bending moment.

Definition of bending moment

It is clear that the bending moment at midspan in this beam is 25.0 kilonewton-metres, and that this quantity can be evaluated by taking moments of all forces to either side of this section. That is to say that the same result would have been obtained if the left-hand half had been covered up, and moments about this point caused by all forces to the right of this section added algebraically.

The term 'bending moment' can therefore be defined as follows: *The bending moment at any section of a beam is the algebraic sum of the moments about that section of all forces to the left or to the right of that section.*

The maximum value of the bending moment, does not, however, occur at midspan in this particular case. If moments of all forces to the left or to the right of the point load are evaluated, the bending moment at this position is seen to equal 33.3 kilonewton-metres. Abbreviating the term 'bending moment' as 'BM', the calculation reads as follows:

BM (from right-hand side) = (3.33 × 10) = 33.3 kilonewton-metres
BM (from left-hand side) = (6.67 × 5) = 33.3 kilonewton-metres

As would be expected from intuitive judgement, the maximum bending moment on a beam loaded with a single point load always coincides with the position of that point load.

Bending moment diagrams

If the value of the bending moment at every point along this beam were calculated according to the principle just defined, these values could all be plotted as the vertical ordinates of a graph. The distances of these points from the left-hand reaction then become the horizontal ordinates of the graph. This representation is shown in Figure 1.5(b), in which the left-hand reaction is the origin of the graph. At this position, the values of both vertical and horizontal ordinates are zero.

The values of the bending moments at midspan and at the point load, 25.0 kN-m and 33.3 kN-m respectively, have already been calculated and can be plotted using a suitable scale. If the bending moments at frequent intervals along the length of the beam are evaluated, sufficient vertical ordinates can be plotted to reveal the shape of the graph. The values of the bending moment at 2.5 metre intervals can be expressed as follows, if moments of forces to the left of each section are taken.

BM at 2.5 metres from R_L = (6.67 × 2.5) = 16.7 kN-m
BM at point load (already calculated)   = 33.3 kN-m
BM at midspan   (already calculated)   = 25.0 kN-m

BM at 10 metres from R_L =
$$(6.67 \times 10) - (10 \times 5)$$
$$= 66.7 - 50 \qquad\qquad\qquad = 16.7 \text{ kN-m}$$
BM at 12.5 metres from R_L =
$$(6.67 \times 12.5) - (10 \times 7.5)$$
$$= 83.3 - 75 \qquad\qquad\qquad = 8.3 \text{ kN-m}$$

The last two bending moment values could have been more easily calculated by taking moments to the right of the section, which are 2.5 and 5 metres from R_R. These would be expressed as:

$$BM \text{ at 5.0 metres from } R_R = (3.33 \times 5) = 16.7 \text{ kN-m}$$
$$BM \text{ at 2.5 metres from } R_R = (3.33 \times 2.5) = 8.3 \text{ kN-m}$$

In cases where the loading pattern on a beam is complicated, it is often advisable to ascertain that the value of the bending moment is the same when moments are taken from either side. If they are not, an error in the arithmetic will have occurred at an earlier stage in the analysis. There would be an erroneous assertion that ΣM does not equal zero, implying that the beam is revolving rather than at rest. If the concepts are clearly understood and the arithmetic correctly performed, the bending moment can be evaluated from the side involving the least amount of calculation.

Now that enough bending moment values have been calculated, they can be plotted as vertical ordinates and joined together as shown in Figure 1.5(b). The resulting shape, which in this case consists of two straight lines, is known as the bending moment diagram. This is simply a graph, therefore, which shows the variation in bending moment along the beam.

A beam carrying a single point load, or for that matter a series of point loads, without any other type of loading, will always yield a bending moment diagram made up of straight lines. But not all bending moment diagrams consist solely of straight lines. This particular graph is a linear function because the bending moments are evaluated using only the first power of any of the terms involved. There are no squared quantities.

A clear understanding of the concept of the bending moment diagram is vital for two reasons. First, the size of a beam, and of many other structural elements, is determined with direct reference to the maximum value of bending moment within the system. Second, the shape of the bending moment diagram is a useful approximate guide to the most appropriate structural form for that element.

Shear force diagrams

It is worth introducing another type of graph at this stage, even though it does not have the same direct relationship to structural form as the bending moment diagram. The shear force at any section of a beam may be defined as *the algebraic sum of the vertical forces to the left or to the right of that section.*

The definition is almost self-explanatory. The loads and reactions to the left of any section can be added up, attributing a positive sign to the upward forces and a negative sign to the downward forces. Using the example of the beam carrying the single point load in Figure 1.5, the graph shown below the bending moment diagram is the shear force diagram. It is simply a representation of the values of the shear force at every section along the beam. All the vertical ordinates between the left-hand reaction and the point load have the same value of 6.67 kN because there are no other loads acting within this zone. This part of the graph is therefore a horizontal line. The same applies to the zone between the point load and the right-hand reaction, in which the shear force is 3.33 kN. The essential feature of this graph is that it crosses the base line, at which the vertical ordinate is zero, at the position of the point load. This position is called the position of zero shear. As can be seen from the vertical alignment of the two graphs, the maximum bending moment also occurs at this position. The relationship between the shear force and bending moment at any section is that the shear force is equal to the slope of the bending moment diagram at that section. The horizontal lines in the shear force diagram thus reflect the constant slope of each corresponding part of the bending moment diagram.

Had the shear force diagram in Figure 1.5(c) been drawn by plotting the shear force values from the right at all the points on the beam, using the same sign convention, the positive and negative values would have been reversed. The distinction between positive and negative shear is not critical to the level of understanding crucial to architects. It is important for engineers confronted with the problem of which way to align diagonal members in trusses, or the diagonal 'shear' reinforcement in reinforced concrete beams. Whilst failure of a beam due to bending is easy to simulate or visualize, failure due to vertical shear is not. This is because the vertical stresses arising from vertical shear forces are always accompanied by horizontal stresses. The failures associated with these stresses will therefore occur on the resultant diagonal plane. It is, in fact, the horizontal shear stresses which have to be carefully examined in those structural elements formed from components with potential planes of separation, such as timber-laminated beams and steel plate girders.

Conic sections

Before examining more complex loading patterns, such as the more commonly occurring uniformly distributed load, a brief introduction to curved lines is called for. Without doubt, the curve which provides the deepest understanding of structural form is the parabola. The parabola will emerge continually in this book as a mathematical concept linking apparently disparate structural forms. It is best introduced graphically as one of a family of four curves known as the conic sections.

Figure 1.6 shows a cone which can be imagined to be cut by four different plane surfaces. If the surface of the cone has a minimal thickness, the horizontal cutting plane will intersect the cone to form a circle. By tilting the plane away from the horizontal, the curve becomes an ellipse. If the cutting plane passes through both the upper and lower halves of the cone, the curves formed are the two branches of a hyperbola. Finally, a cutting plane parallel to one of the sides of the cone will yield a parabola.

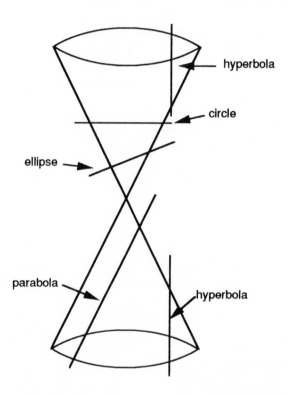

Figure 1.6 *The conic sections*

This relationship between a cone and these four curves was first established by the Greek mathematician Apollonius during the third century BC. Yet it was not until the sixteenth and seventeenth centuries that the connection between these curves and the laws of physics was recognized. Johannes Kepler (1571-1630), for example, established that the planets in our solar system travelled in elliptical paths around the sun. It had earlier been suggested by Copernicus (1473–1543) that the planets traced out circular orbits, the idea of the earth being situated at the centre of the known universe having been discarded. There is a strong similarity between the two curves. Whereas the ellipse has two generating points known as foci, the circle has only one, the centre. An ellipse can in fact be constructed within two concentric circles, the major and minor axes being the diameters of the larger and smaller circles respectively.

The similarities between the other two curves and the ellipse are not so immediately apparent unless explored in terms of their critical internal dimensions. Apollonius was able to differentiate between them by expressing them as variants within a particular geometrical construction. It is perhaps simpler to understand these differences by looking at the meanings inherent in the names of these curves in their adopted English usage. 'Parabole' means, in rhetoric, a simile or metaphor. A speaker resorting to the use of 'hyperbole' is guilty of wild exaggeration, whilst an 'ellipsis' conveys the impression of words omitted and implied.

The parabola

Apollonius, although investing this particular conic section with a suggestion of truth in speech, was probably unaware of its potential for describing the physical world. It could be argued that the essential clue to the link between the parabola and an ideal structural form was provided by the French mathematician and philosopher René Descartes (1596–1650). Cartesian axes and coordinates emanate from his name, and are the basis of graphs such as that shown in Figure 1.7.

Using the algebraic symbols x and y as descriptions of the horizontal and vertical axes respectively, a graph can be drawn to represent an equation in which the value of y at any point depends on the value of x. Thus y is a function of x, and each point on the graph is identified by a pair of Cartesian coordinates x and y. The curve in Figure 1.7 is the graph of the equation

$$y = x^2$$

which is the fundamental equation of the parabola. To dispel the erroneous idea that all parabolas have this equation, it is necessary to state the general relationship between x and y in a parabolic curve which is:

$$y = Ax^2 + Bx + C$$

The coefficient A tells just how many x squareds are needed to make up y. The coefficients B, applied to x, and C define the exact contours and the location of the parabolic curve in relation to the Cartesian axes, and are both zero in the symmetrical arrangement in Figure 1.7.

The coordinates of this curve can therefore be located by simply squaring the values of x as follows:

x	y
1	1
2	4
3	9
4	16
5	25

and so on towards infinity. The cutting plane which generated the parabola lay parallel to the surface of this cone, which also extends into the infinite.

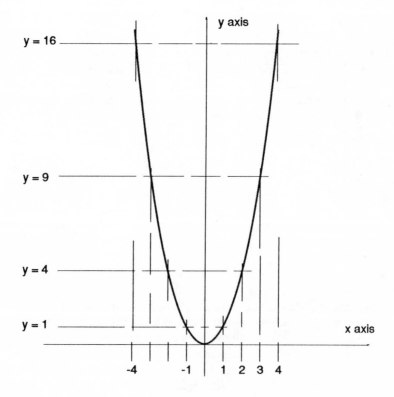

Figure 1.7   *Graph of y = x²*

The parabola and the uniformly distributed load

Earlier sections demonstrated the way in which the bending moment diagram could be constructed for a beam carrying a single point load using the principles of elementary statics. These same principles can be applied in exactly the same manner to the problem of a load spread evenly along the length of a beam, as shown in Figure 1.8. Such a load is known as a uniformly distributed load, usually referred to by the abbreviation 'UDL'.

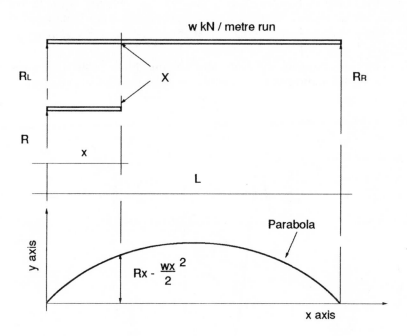

Figure 1.8 *Parabolic bending moment diagram*

Most loading patterns encountered in architecture do in fact conform to this system, since dead and live loadings from floors and roofs will impose forces on supporting structural elements in this manner. The symbol w denotes the magnitude of the UDL in kilonewtons on every metre length of the beam, being expressed therefore as 'w kN/metre run'.

The bending moment at any section 'X' of this beam, distance x metres from the left-hand support, can be written down using the definition given earlier. Thus, taking moments about X of all forces to the left of the section, calling the left-hand reaction simply R in this case,

$$BM \text{ at } X = Rx - wx. \, x/2$$

The first term is the clockwise moment about X, and the moment in the second term, being anti-clockwise, attracts a minus sign. This second term needs some explanation, since it involves the moment of part of a uniformly distributed load. As far as its turning effect, or moment is concerned, any force will act through its centre of gravity. The centre of gravity of a uniformly distributed load, whether the entire load or only a part of it is being considered, will clearly be at its mid-point. For that part of the UDL to the left of X, the centre of gravity will be $x/2$ metres from the point X, at which the magnitude of its moment is required. The moment about X of this partial uniformly distributed load is therefore the force wx multiplied by the distance $x/2$. This can be written more simply by expressing the multiplication of x by itself as a squared quantity, i.e. x^2.

The expression for the bending moment at any point X at a distance x from the left-hand reaction can therefore be simplified to:

$$BM \text{ at } X = Rx - wx^2/2$$

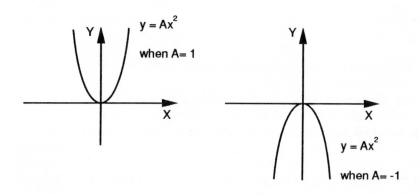

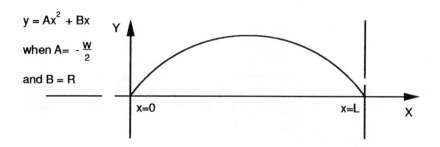

Figure 1.9 *Transformation of parabola*

By rearranging this expression into the form of the general equation for the parabola, i.e.

$$y = Ax^2 + Bx + C, \quad \text{this becomes}$$
$$BM \text{ at } X = -wx^2/2 + Rx$$

The coefficient A is $-w/2$. B is R and C is zero. If all such vertical ordinates were plotted as a graph, the resulting curve would be the parabola shown in Figure 1.8.

The difference between this parabola and that described in the previous section is that the negative value of A and the introduction of B has transformed the curve through the stages shown in Figure 1.9. The negative A has turned the parabola upside down. The non-zero coefficient B has modified its shape and moved its closed portion into the zone bounded by the positive lengths of the axes x and y.

When x is equal to the span of the beam L, the right-hand side of the parabola will cross the x axis. When x equals zero, the left-hand side of the parabola intersects the origin of the graph, that is where the two axes meet. At both of these positions, y is equal to zero. This is to be expected, since the bending moment at each support to the beam must be zero.

The hyperbola and its asymptotes

The appearance of the hyperbola in structure lies in the formation of curved surfaces using straight lines. The surface known as the hyperbolic paraboloid will be explored in Chapter 8. This is formed by translating a downward curving parabola onto an upward curving parabola, as shown in Figure 1.10(a). Any horizontal section through this construction will reveal the two branches of a hyperbola. If vertical sections through a particular point were taken whilst rotating through 360°, the curvature would disappear twice as it changed from an upward to a downward direction and back again. The shape resembles that of a saddle, and occurs in nature in the topography of a mountain pass. The top of the pass is lower than the mountain range which it traverses, and relates to the 'saddle point' on a hyperbolic paraboloid.

The surface shown in Figure 1.10(b), familiar in the cooling towers of power stations, can also be generated by straight lines. The two vertical curves forming the boundaries of the elevated form are again the branches of a hyperbola. This form is known as a hyperboloid of revolution.

The equation of the hyperbola is:

$$x^2/A^2 - y^2/B^2 = 1$$

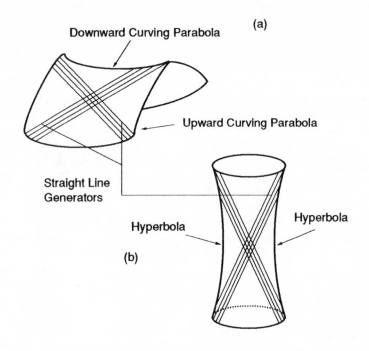

Figure 1.10   *Surfaces generated by straight lines*

where *A* and *B* are constants. Referring to Figure 1.6, a hyperbola consists of two curves, one in each of the upper and lower portions of the cone. There exists within the space between the two curves a pair of straight lines which intersect at the origin of the graph, and approach, but never quite touch, the curves of the hyperbola. These are called 'asymptotes'. An asymptote, defined as a line that continually approaches a curve but never meets it, is not peculiar to the hyperbola. It is a useful mathematical aid to the understanding of the relationship between the geometry of an arch or a suspension cable and the horizontal thrust generated in these forms.

The Ellipse

The ellipse is the familiar oval shape, the equation for which is:

$$x^2/A^2 + y^2/B^2 = 1$$

where *A* and *B,* as in the case of the hyperbola, are constants. The ellipse has a horizontal major axis and a vertical minor axis when $A > B$. *A* and *B* are one half of the lengths of the major and minor axes respectively. Given the information of the length of the major axis and the position of the foci, an ellipse can easily be drawn. If each end of a piece of string of the same length as the major axis is tied to each focus, the curve traced out by a pen touching the taut string will be an ellipse. This property makes the ellipse an easy curve to set out, and has, not surprisingly, been used as an arch form in bridges and building structures.

There is not, however, the same valid structural logic underlying the ellipse as there is in the parabola. The relevance of the ellipse in physics lies principally with the elliptical orbits of electrons in an atom, in which the nucleus coincides with one of the foci.

The circle

If the two foci of an ellipse occupy the same position in space, the major and the minor axes are the same length, and the curve becomes a circle. The single axis is the diameter. The circle is clearly a unique form of an ellipse. The constants *A* and *B* are equal, and the equation of a circle becomes:

$$x^2 + y^2 = R^2$$

in which *R* is one half of the diameter, i.e. the radius.

This is again an easy curve to set out, as witnessed by arches constructed in the form of semi-circles or segments of circles. As with the ellipse, ease of construction is not necessarily compatible with perfection in structural form. This theme will be pursued in Chapters 5 and 6.

The chief interest in the circle as a structural element is in its capacity to resist uniformly distributed loads applied perpendicular to the tangents at every point on the perimeter in the plane of the circle. A tangent is a line that touches a curve without passing through it. All tangents to a circle are at right angles to the radius. A force applied normal, or perpendicular, to a tangent can therefore be resolved by triangulation into the tangents to the immediately adjacent parts of the circle. If the force normal to every tangent is equal, that is if the load is uniformly distributed, the circle will respond by accepting a constant force, without any bending moments. This condition is known as ring tension or ring compression, depending on whether the tangential forces were pushing inwards or pulling outwards.

General formulae for bending moment

It is appropriate at this stage to derive the expressions for the maximum bending moment on a simply supported beam for the two most commonly occurring loading conditions. Once derived and stated in terms of the algebraic symbols representing load in units of kilonewtons and span in units of metres, they can be easily committed to memory, and applied with informed judgement.

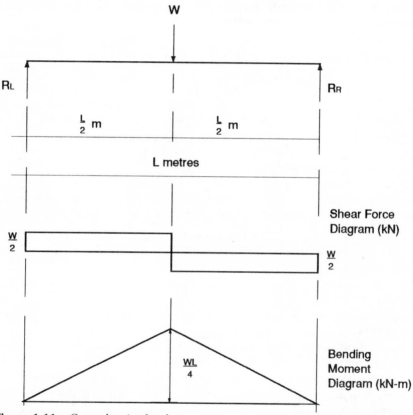

Figure 1.11   *Central point load*

Central point load

Earlier sections explored the way in which a bending moment diagram could be drawn after having used the laws of statics to calculate reactions and values of bending moment at distinct points along the beam. The point load in Figure 1.5 was not at mid-span, so that the magnitudes of the two reactions differed. Figure 1.11 depicts a load of W kilonewtons at the mid-span position of a beam with a span of L metres. By using the principle of moments to determine the reactions, R_L and R_R can each be shown to be $W/2$ kN. This division of the load into two equal reactions is also immediately apparent from the symmetry of the system. The bending moment at mid-span will then have a value of

$$(W/2 \times L/2) = WL/4 \text{ kilonewton-metres, or}$$
$$WL/4 \text{ kN–m}$$

The shear force diagram drawn above the bending moment diagram in Figure 1.11 also confirms the intuitive judgement that the position of maximum bending moment will be at mid-span.

Uniformly distributed load (UDL)

Using the symbol 'w' introduced earlier to denote the magnitude of a UDL, Figure 1.12 shows the loading, shear force diagram and bending moment diagram for this load over a span of L metres. The system, like the beam with the central point load, is symmetrical. Each reaction will therefore be equal to one half of the total load. The shear force diagram consists of a sloping line because the loading, and therefore the algebraic sum of the vertical forces, or shear force, changes constantly along the length of the beam. The identical position of zero shear force and maximum bending moment is again confirmed in Figure 1.12.

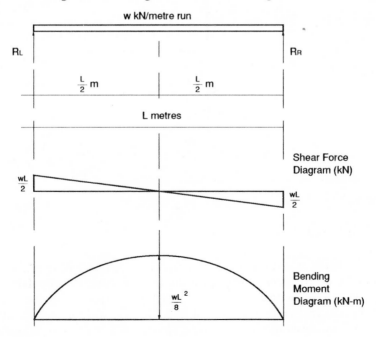

Figure 1.12   *Uniformly distributed load*

The shape of the bending moment diagram was shown earlier to be a parabola. Since the total load on this beam is w kN/metre multiplied by L metres which equals wL kN, each reaction is equal to $wL/2$ kN.

Into the expression for the bending moment on a beam carrying a UDL can be substituted this value for R and the value for x at mid-span of $L/2$.

The maximum bending moment, also the maximum height of the parabola, is therefore

$$
\begin{aligned}
Rx &- wx^2/2 \\
&= wL/2.L/2 - w.(L/2)^2/2 \\
&= wL^2/4 - w.L/2.L/2.1/2 \\
&= wL^2/4 - wL^2/8 \\
&= wL^2 \ (1/4 - 1/8) \\
&= wL^2/8 \ \text{kN} - \text{m}
\end{aligned}
$$

This result is often expressed in terms of the total load W, so that the maximum bending moment is a function of this total load on the beam rather than the load per unit length. Since W in this case will be equal to wL, it can be substituted in the expression derived above to give a value of $WL/8$ kN – m.

By comparing this expression with that for the maximum value of bending moment under a central point load, i.e. $WL/4$, it can be seen that for the same total load on a simply supported beam of the same span, the maximum bending moment is exactly double the value when it is concentrated at a single point than it is when evenly distributed along the beam.

Chapter 2

MATERIALS AND FORM

Every building has its own spatial identity. Whether the occupant lingers in one part of a building or moves through a succession of enclosed spaces, the architectural experience is partially informed by a response to shapes and volumes. The identity of a building is also dependent on the properties of the materials from which it is constructed. If the material selected by the architect has mechanical properties compatible with the strength requirements of the structural elements, while possessing qualities of texture and colour conducive to the desired atmosphere of a space, the decision to choose that material could reasonably be said to have been a wise one.

Whether the material used for the structure and the enclosing elements of a building is expressed visually, or whether decorative or protective surface finishes are used, is a matter for the architect to decide. From considerations of structure, those material properties which directly affect the transmission of forces to the building's foundations must be clearly understood.

There are certain building types which appear by their very nature to demand to be built in a particular material. A case in point is a house, in which bricks for the walls and timber for the floors and roof trusses may seem an obvious choice. Yet, as a result of the ideas surrounding the Modern Movement earlier this century, reinforced concrete became associated with some of the more striking examples of European domestic architecture in the 1920s and 1930s. More recently, timber-framed walls have frequently been preferred to brickwork on grounds of economy and speed of erection. Irrespective of whether the decision is based on the market place or on prevailing architectural fashion, the materials used in any building must possess qualities of strength and durability compatible with the loads to which the structural elements will be subjected.

States of stress

The strength of a structural element is a measure of the load which it is safely able to support. This capacity depends on two parameters – the material from which the element is made, and its cross-sectional area. The relationship of load, or force, to cross-sectional area is expressed as a measurement of stress.

Stress is defined as force per unit area. A force of 200 Newtons applied to a cross section measuring 5mm by 5mm, an area of 25mm², will produce a stress of

$$200/25 = 8 \text{ Newtons per square millimetre, or}$$
$$8 \text{ N/mm}^2$$

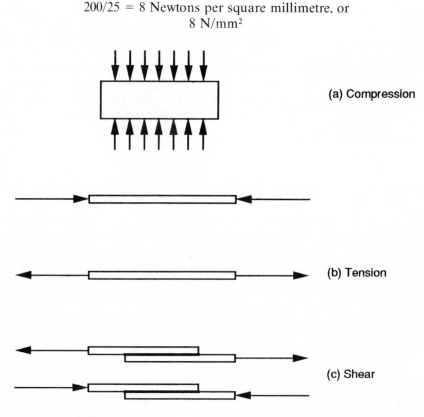

(a) Compression

(b) Tension

(c) Shear

Figure 2.1   *States of stress*

The three basic forms of stress are compressive, tensile and shear. The first two are differentiated by the direction in which the force is applied. The third can arise from forces applied in either direction. Figure 2.1 shows these three states, in which each pair of vector arrows represents action and reaction in equilibrium. The shear stress in Figure 2.1(c) is applied to the membrane connecting the two components together. The nature and magnitude of this stress will be the same whether the force vectors are pushing inwards or pulling outwards, although in the first case the joined components will be in a state of tensile stress, and in the second case they will be in a state of compressive stress. This type of connection occurs in glued timber joints and welded steel joints. The magnitude of the shear stress in this case will depend on the area of the connecting membrane in the plane of the forces, and not on the cross-sectional area of the connected parts.

Unfortunately, states of stress do not exist in total isolation from one another. A steel bar, for example, tested to destruction under an externally applied tensile force does not fail on a cross-sectional plane. The two portions of the bar part company at an angle of $45°$ to the longitudinal axis. The failure is a result of the shear stress exceeding a certain value. A cast iron bar, on the other hand, will, under the same tensile force, fail in the expected manner, that is at right angles to the longitudinal axis.

The failing, or ultimate, stress of a material, however, is usually defined in terms of the direction of the force which causes the failure, rather than the orientation of the plane of separation. The ultimate stress of mild steel, for example, is about 450 N/mm^2 and refers to a force of 450 Newtons applied to every square millimetre of steel perpendicular to the axis of the test specimen.

Compressive and tensile force vectors

Given the relationship between force and stress, it is important at this stage to establish a sign convention depicting resistance to axial forces, that is forces applied along the axis of a structural member. Figure 2.2 shows the vector notation for externally applied and internally resisted compressive and tensile forces.

In Figure 2.2(a), the action and reaction are pushing inwards, and therefore produce a state of compression in the structural member. The state of compression always causes a shortening of the member, which responds by pushing outwards. This outward push constitutes the internal reaction of the member, and is depicted by the two arrows pointing in opposite directions. Each pair of vector arrows at the ends is mutually opposed, thus modelling the state of equilibrium at each end of the member. These pairs are ringed together. The four arrows therefore represent states of external and internal equilibrium throughout the

member. The extreme left-hand arrow is the external action, the extreme
right-hand arrow is the external reaction, and the two internal arrows are
the internal reaction. The important feature of this convention is that the
internal arrows, which in complex structural forms consisting of a large
number of members are used to read the type of force and therefore the
state of stress, reveal the manner in which the member is responding to
externally applied forces.

Figure 2.2(b) relates to externally applied forces pulling outwards from
the structural member, thus putting that member into a state of tension.
By the same logic used in the previous paragraph, the external force
vectors are outwardly directed, and the internal vector arrows are shown
pulling inwards.

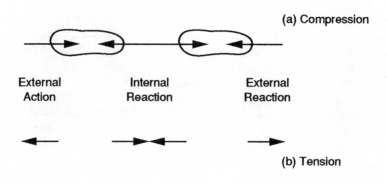

(a) Compression

External   Internal   External
Action   Reaction   Reaction

(b) Tension

Figure 2.2   *Action and reaction*

The notation used in this sign convention is in keeping with the way in
which material forms, or human forms for that matter, respond to forces
imposed on them from an external source. The fibres of a rope with a
weight on one end will transfer that weight to the fixing at the other end
by pulling inwards. Every particle of the Corinthian column in Plate 1
will push outwards in its endeavour to transfer the load imposed at the
head of the column to its base. Although the notion of force vectors was
not available to Greek and Roman architects, the need to express the
transfer of loads at the extremities of compression members was a salient
feature of the language of Classicism.

Plate 1   *Corinthian column*

Deformation and strain

It is a fundamental law of physics that all forces imposed on material objects cause deformations. An elastic band when stretched becomes perceptibly longer. The difference between its original length and the length in its stretched condition is known as the deformation. The ratio of the deformation to the original length is called the strain. Since the numerator and the denominator are expressed in the same units, strain has no dimensions – it is a ratio. It is a common error in everyday language to use the words 'stress' and 'strain' as though they were interchangeable and synonymous. If a further ratio is written down, that is by dividing stress by strain, the property that results is an indicator of the resistance of the material to deformation under load. An elastic band offers little resistance to stretching when subjected to a small tensile force. A steel wire having the same cross-sectional area and subjected to the same small tensile force will, on the other hand, show a deformation which is so small as to be imperceptible to the naked eye. The stress to strain ratio will therefore be much higher for the steel wire than for the elastic band.

Modulus of elasticity

The ratio of stress to strain is known as the modulus of elasticity of the material. The Latin word from which 'modulus' is derived means a measure, and it is defined in the English dictionary as 'a constant multiplier or coefficient'. The stress to strain ratio of a material is a coefficient which has a low value for extensible matter such as rubber and a much higher value for metals.

The generally accepted symbol for the modulus of elasticity is E. Since E is obtained by dividing a value of stress by a ratio, i.e. strain, its units will be the same as those of stress, e.g. Newtons per square millimetre.

Given that

$$\frac{\text{STRESS}}{\text{STRAIN}} = E \quad \text{and}$$

$$\text{STRAIN} = \frac{\text{DEFORMATION}}{\text{ORIGINAL LENGTH}}$$

the ratio strain will equal unity when the stressed material doubles its length. This happens when the deformation and the original length are equal, a condition which can easily be imagined or simulated for the elastic band. Feeding this value of strain equal to one into the first expression, E becomes the level of stress required to double the length of the test specimen. Although this is a convenient mental reference for the meaning of E, structural materials will have failed well before this point is reached. The stress to strain relationship is valid only within that range where load is proportional to deformation. This mode of behaviour is called elasticity, and was first identified by Robert Hooke (1635–1703), a near contemporary, but not by most accounts a close friend, of Isaac Newton. Hooke's Law applies within the elastic range of a material, but beyond that, the deformation, and therefore the strain, start to increase without the addition of load as eventual failure is approached.

Robert Hooke and architecture

Inasmuch as those blessed with the most creative minds have made contributions in both the arts and the sciences, it may not be thought too idle to pause briefly to examine Hooke's contribution to architecture. Hooke, in his capacity as one of the City of London's surveyors, had been associated with Sir Christopher Wren (1632–1723) in the building of The Monument and Wren's city churches. Hooke has also been credited with the design of several buildings in and around London, including Willen Church in Buckinghamshire, although his aesthetic sensibility has been questioned (Summerson, 1978, p.60). To assert that Hooke

Plate 2   *St. Mary Woolnoth*

influenced the elevational treatment in the designs of his contemporaries is pure speculation. There is, however, in the rusticated masonry in the church of St Mary Woolnoth (Plate 2), one of the London churches of Nicholas Hawksmoor (1661–1736), a pupil of Wren, the suggestion of a spring vibrating under a compressive force. This church was built between 1716 and 1724, about two decades after Hooke's death. Although there is little evidence that Hooke's discoveries in physics influenced Hawksmoor's ideas on design, the connection is a tempting one to make.

Bending stresses

The bending moments described in Chapter 1 also produce stresses in the structural members to which they are applied. Just as a moment is an effect caused by the displacement of the planes in which forces are acting, bending stresses are equivalent to tensile and compressive stresses acting over separate areas within the same section. These stresses, when multiplied by the areas over which they are applied, produce tensile and compressive forces respectively. This is a consequence of the relationship:

$$\text{STRESS} = \frac{\text{FORCE,}}{\text{AREA}} \quad \text{from which}$$

$$\text{FORCE} = \text{STRESS} \times \text{AREA}$$

The existence of tensile and compressive stresses in a bent form can be visualized with reference to the bending of a piece of a material with a low E value, such as a piece of india rubber. As shown in Figure 2.3, one surface of the test piece has become noticeably shorter, and the opposite surface noticeably longer. The deformed shape of the rubber has been brought about by applying a moment directly to it, this action being achieved through the rotation of either end in opposite directions using thumbs and forefingers. The clockwise and anti-clockwise moment vectors of equal magnitude M produce a constant moment throughout the length of the specimen. The bending moment diagram for this system can be drawn in exactly the same way as the bending moment diagram for a system of vertical forces. Here, the bending moment increases from zero to M at the left-hand end, stays at this value for the length of the specimen, and, following the application of the vector $-M$ at the right-hand end, returns to zero.

In passing from a state of compressive strain at the top surface of the rubber to a state of tensile strain at the bottom surface, there must be, by deductive logic, a horizontal plane on which there is no strain at all. This plane is called the neutral axis, above which the compressive strain increases to a maximum at the top surface and below which the tensile strain increases to a maximum at the bottom surface. The increase is

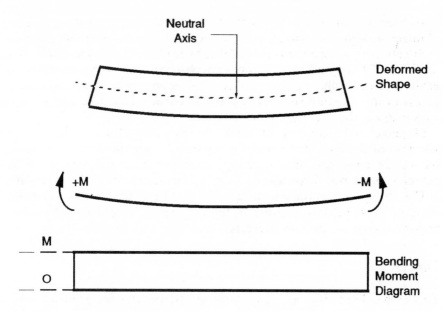

Figure 2.3 *Beam subjected to constant bending moment*

linear in both directions. The top and bottom surfaces of a bent structural element are known as the extreme fibres in compression and tension respectively.

Since stress is proportional to strain in an elastic material, the variation in stress throughout the depth of the vertical section of rubber will be identical to the variation in strain.

This mode of behaviour is conditional on the premise that the material in this bent form obeys Hooke's Law. There is little doubt surrounding the elasticity of rubber. The degree to which this theory can be legitimately applied to the more familiar construction materials is dependent on their behaviour at various stages of loading. The elastic theory of bending, as it has come to be known, has until quite recently been the basis of design for all structural materials, and when applied with judgement informed by a clear understanding of structural behaviour, has always resulted in safe structures.

Structural materials

Descriptions of the atomic structures and the processes of manufacture associated with building materials are not essential to an understanding of structural form. What follows is a brief review of their main characteristics as they influence choice of structural form, and the visual qualities with which their adoption can endow a building. As well as possessing variations in the more obvious structural properties such as ultimate strength and modulus of elasticity, structural materials differ in their capacity to resist the action of fire, their changes in appearance over a period of time, and their ability to withstand the ravages of rain and environmental pollution. Availability is an important consideration, as is access to the building site of the plant and machinery necessary for the conversion of the material to a structure.

Elegance and economy of form are conditioned not only by the selection of the correct material for a particular architectural expression, but by an awareness that a change in material at a certain level, or between spaces with markedly different volumes, may be a wise decision.

The ability of a particular material to resist one or more of the basic states of stress discussed earlier in this chapter is, nevertheless, the most important consideration in structure.

Timber

Timber has been used as a structural material ever since man acquired the tools for cutting down trees and converting them into shapes suitable for enclosing primitive shelters. In its felled condition, timber has a high moisture content, which needs to be reduced by seasoning before being used in modern structures. A reduction in moisture content, which is defined as the percentage ratio of contained water to the oven dry weight of wood substance, increases the strength properties considerably. Structural timber is generally regarded as dry when the moisture content is below 18 per cent.

Although wood tissue has a weight density of about 15 kN/m³, the weight density of all species of timber is much lower, because of the cell voids present in the botanical structure. The values are wide-ranging even for hardwoods, varying from about 1 kN/m³ for the balsa wood used widely in model making to about 10.6 kN/m³ for greenheart. The deciduous hardwoods are used very little in the United Kingdom for structural purposes. The coniferous softwoods most commonly used have weight densities between 5.4 kN/m³ for Western Hemlock and 7.2 kN/m³ for Pitch Pine.

Timber is the lightest of the structural materials. As such, the self weight of timber structures makes only a small contribution to the bending moments of spanning elements such as beams and trusses. Provided that the dead load which they support is also constructed of timber or an equally light form of decking, the superimposed loads are usually the most critical. One would not in the normal course of events propose a scheme in which timber joists or beams were asked to support a reinforced concrete floor deck. Lightweight structures are compatible only with the lightest possible loads.

Being suitable for sawing into long members of small cross-section, timber is best thought of as a material from which one-way spanning forms are generated. This implies a system of parallel members such as floor joists spanning in one direction, supporting a series of parallel secondary members such as floor boards spanning perpendicular to them. Although curved membranes have been constructed in which thin timber sections at right angles to each other have been connected at their intersections to form a grid, timber is not a natural two-way spanning material. It is not possible to make two identical timber sections pass through each other in the same plane in an organic sense, as it is with reinforced concrete.

Since the wood tissue forms long tubular cells, timber is resistant to both compressive and tensile stresses. The angle which these cells make with the longitudinal axis of a piece of timber is known as the slope of the grain. The smaller the slope, the greater is the compressive and tensile strength. The compressive strength of timber perpendicular to the grain is much less. This is not a great structural disadvantage, although the capacity of a timber sole plate to support the load from the vertical studs in a timber-framed wall is determined by this property.

Strength in both tension and compression make timber an ideal material for resistance to bending moments. This attribute is only possessed, however, by virtue of the resistance of timber to shear stresses parallel to the grain. Without this shear strength, separation along the interfaces of the cell walls could occur under the action of bending moments caused by loads perpendicular to the axis of the member. The relationship between shear force and bending moment was explained in Chapter 1, in which the shear force at any section was defined as the slope of the bending moment diagram.

Since timber is a material whose existence is wholly dependent on the processes of natural growth, its condition is organic. It is therefore combustible. The fact that it will burn when exposed to fire, however, does not mean that it is devoid of fire resistance. The burned fibres of the timber on the outside of an element subjected to fire provide a certain degree of thermal insulation between the fire source and the timber inside this peripheral zone. This phenomenon is acknowledged in the current British Standard Code of Practice, BS 5268. The design philosophy is based on the prediction of the size of the timber section remaining after burning has taken place for a particular time interval. The speed of burning is known as the charring rate, which decreases with the density of the species. The prediction of the fire resistance period of a timber structural element is, however, rather more involved than this in that the quality of the connections into adjacent members has to be considered.

Concrete

The principal determinant of the way in which concrete is used in structures is its strength in compression and its very low tensile strength. Where tensile stresses occur in structural concrete, steel is provided in sufficient quantity, and in the appropriate positions, to resist the corresponding tensile forces, giving rise to the terminology of reinforced concrete. Only in those structural elements where very small diagonal tensile stresses, caused by the resultants of vertical and horizontal shear forces, occur, such as slabs or lintels, is the limited tensile strength of concrete acknowledged.

The contrast between the compressive and tensile strengths of concrete, even without the equipment necessary for testing, is predictable from a knowledge of its composition. The concrete familiar to us in modern structures dates back to the invention of Portland cement by Joseph Aspdin, a bricklayer from Leeds, in 1824. Graded stones and sand, known as aggregates, when mixed with hydrated cement, result in a composition that is instinctively felt to be more resistant to external forces pushing inwards rather than pulling outwards. It is incidentally impossible to induce a pure failure in compression in a concrete specimen, as witnessed by the diagonal planes of separation in test cubes and cylinders. The ultimate failure is actually caused by the tensile stress on these planes exceeding the tensile strength of the concrete.

The compressive strength of concrete varies with the mix proportions and the amount of water used. Concrete becomes stronger with age – an age factor is often used to assess the capacity of reinforced structures to accept the increased loading that may occur following the change of use of a part of a building. For design purposes, the stress at failure in concrete made with ordinary Portland cement 28 days after casting is taken as the criterion of strength. Most concrete mixes used in architecture have cube crushing strengths of between 20 and 25 Newtons per square millimetre.

Concrete cast with normal aggregates has a weight density of between 20 and 24 kN/m^3. For concrete cast with lightweight aggregates such as pulverized fuel ash, the figure is around 16 kN/m^3. Although the latter is often referred to as lightweight concrete, the adjective is a misleading one when it is remembered that the density of commercial softwoods is about 6 kN/m^3. Concrete made with lightweight aggregates has a greater strength to weight ratio than concrete made with normal aggregates with the same mix proportions. This advantage is partially offset by the lower E value of the lighter material, sometimes resulting in deeper structural members to limit deflections. The weight reduction is nevertheless of particular advantage when the foundation design demands the smallest possible dead loads from a reinforced concrete superstructure.

Reinforced concrete

From the days of its early use in buildings, it has been customary to use reinforced concrete in a trabeated form, in which a one-way spanning mode is employed. In other words, the relationship between a floor slab and its supporting beams in reinforced concrete is analogous to that between a floor board and a joist in timber. The realization that the placing of steel bars along the lines of induced tensile force in a concrete member, thus creating a structural element capable of resisting bending moments, led to reinforced concrete being used to form linear elements in the same way that one would use timber.

Reinforced concrete in its pre-cast form does certainly behave as a series of parallel one-way spanning elements, because there is little or no interaction between adjacent members. When cast in moulds on the site, or *in situ,* however, reinforced concrete assumes the geometry of a plane rather than of a line. Provided that the arrangement of steel reinforcing bars permits reactions to tensile forces in two horizontal directions at right angles to each other, more succinctly called orthogonal directions, the structural element can be regarded as capable of spanning two ways. Because concrete in its hardened state will take the shape of the formwork into which it is poured, its sculptural potential is unlimited. Provided, therefore, that the critical dimensions of the reinforced concrete members are compatible with the forces and bending moments imposed on them, there exists an almost unlimited choice of structural forms available to the architect. The two-way spanning mode is a more natural behaviour pattern for reinforced concrete than a one-way system. Chapter 8 explores the theory underlying two-way spanning in both reinforced concrete and structural steel.

A further corollary of the nature of reinforced concrete is the property of continuity between one structural element and another. The casting of a floor or roof slab integrally with its supporting beam increases the strength of that beam beyond that of its immediate rectangular section. This composite form is referred to, because of its shape, as a 'T' beam. Also, the continuity between a column and a beam in reinforced concrete creates a condition known as 'fixity', in which bending moments can be transferred between the two elements. This is particularly important when the structural system has to act in the manner of a portal frame, in which the horizontal forces are resisted directly by the framed structure rather than by the enclosing and internal walls. Portal frame action is dealt with at some length in Chapter 5.

Unreinforced masonry

The absence of any significant tensile strength in unreinforced concrete is a property shared by natural stone and manufactured bricks and concrete blocks. This range of materials is collectively known as unreinforced masonry. Lack of tensile strength restricts these materials to those structural elements in which only compressive stresses arise. They are clearly suitable for vertical load-bearing elements such as walls, piers and columns. Some arch configurations, discussed in Chapter 5, are also geometrically compatible with stone, brick and block although, as proved in that chapter, it is not true to say that all arches are permanently in compression.

Brickwork and blockwork have been used in a reinforced condition by threading longitudinal reinforcement through holes cast into these components, inducing a mode of behaviour analogous to that of reinforced concrete. This system has been common in retaining wall construction for a number of years. More recently it has been used with considerable ingenuity to create the impression in some buildings of bricks spanning horizontally over large openings, an effect traditionally achieved by the use of facing bricks concealing a steel or reinforced concrete beam.

The craft of masonry involves the formation of structural elements by an assembly of smaller components, using mortar at the joints to maintain the continuity of cross-sectional area. Mortar, whether formed with cement according to current practice, or lime as was the custom from Roman times until the invention of cement, must also be considered as devoid of tensile strength. It is not an adhesive in the sense that it provides a continuity of the limited tensile strength of the masonry. The thickness of any masonry wall must be great enough to ensure that no tensile stresses occur under any conditions of loading. One form of walling where no mortar is used is dry stone walling, a craft perfected in upland farming areas where workable stone such as limestone is available in plentiful supply. This is one of the few examples where the thickness of a wall can be perceived. In buildings, only the inside and the outside surfaces of a wall are visible, though clearly not at the same instant in time. Its thickness remains concealed to the naked eye. Masonry forms in structure are not immediately legible.

There is an enormous variation in both density and the compressive strength of individual components. The weight density of a common or a facing brick may be as low as 14 kN/m^3 and that of an engineering brick as high as 25 kN/m^3, with corresponding ultimate crushing strengths as low as 3.5 N/mm^2 and as high as 140 N/mm^2 respectively.

Concrete blocks for load-bearing use are required to have a minimum density of 15 kN/m^3, whilst the average ultimate crushing strengths can range between 3.5 N/mm^2 and 35 N/mm^2.

Although quarried stone is used only in very small quantities as a load-bearing material in the United Kingdom as compared with clay bricks and concrete blocks, permission to build in National Parks, Areas of Outstanding Natural Beauty and in some historic city centres may be conditional upon the use of the vernacular material. The heaviest and strongest building stone is granite, with a weight density of up to 27.5 kN/m³ and a crushing strength as high as 147 N/mm². The corresponding values for limestone, to which classification Portland and Bath .stones belong, are 23.5 kN/m³ and 65 N/mm² respectively.

Steel

It is ironical that steel always has some part to play in the design of lightweight structures, yet is the heaviest of the materials used for building. The weight density of steel is 72 kN/m³, about three times that of concrete and six times that of commercial softwood. Partly for this reason, steel is used in a solid form only for components whose cross-sectional area is very small, such as reinforcing bars, prestressing wire or ties in trussed members. Steel sections used as beams and columns are formed by rolling the reheated material into shapes in which the steel occupies the most highly stressed positions within an enclosing rectangle, square or circle. In 'I' shaped and hollow sections there is more vacant space than solid steel. The reason for the adoption of these particular shapes is explained in Chapters 3 and 4.

Steel sections are always specified by the enclosing dimensions of a section and the mass per metre length of that section. For example, a steel column 254 mm deep and 254 mm wide is rolled with varying thicknesses of the vertical and horizontal portions, called the web and flanges respectively. The heaviest section is specified as

$$254 \times 254 \times 167 \text{ kg/metre}$$

meaning that each metre length of that section weighs

$$(167 \times 9.81) = 1638 \text{ Newtons} = 1.638 \text{ kN.}$$

The weight of a one metre long solid piece of steel with a cross-sectional area of 254 mm by 254 mm is, incidentally,

$$(0.254 \text{ metres} \times 0.254 \text{ metres} \times 1 \text{ metre} \times 72 \text{ kN/m}^3) = 4.64 \text{ kN,}$$

nearly three times that of the heaviest 'I' section.

The strength properties of steel are guaranteed by the quality control system inherent in the production process. There is not the wide variation in strength that exists with masonry and timber. Steel is a perfectly elastic material up to a stress level known as the yield point, beyond which strain increases at a greater rate than stress. This range in which Hooke's Law no longer applies is called the plastic range. The various grades of steel are specified by both their yield points and their ultimate tensile strengths. Mild steel, which has a yield point of 230 N/mm², ultimately fails in tension at about 450 N/mm².

When compared with the other materials dealt with in this chapter, the order of magnitude of the strength of steel is clearly much greater than that of timber and masonry. The disadvantages of the material lie in its susceptibility to corrosion and its limited resistance to fire.

Until quite recently, steel beams and columns were required to be encased in concrete in order to provide a reasonable period of fire resistance. It is now accepted, however, that lightweight cladding carefully applied can insulate the surface of the steel, thus preventing the rise in temperature which would cause deformations and reduce the structural integrity of the section. In some recent buildings, the presence of water circulating within hollow sections, which conducts the heat away from the inner surface of the steel almost as soon as it is absorbed from the outside, near to the source of the fire, has permitted designs in which the steel structure can be totally expressed. The fire resistance of the steel structure of Bush Lane House in London (Plate 3) has been achieved in this way.

There is another strange irony in that timber, an organic material and therefore combustible, does possess a measure of fire resistance by virtue of the charring process whilst steel, a metal and therefore incombustible, has virtually no fire resistance.

Structural joints

It is not too much of a simplification to assert that the choice of some structural materials avoids problems of detailing whilst the choice of others creates them. This is not to say that a minimal time spent on details should be the sole aim when examining the relative merits of the various materials at an architect's disposal, but it is as well to recognize at an early stage in design that adjacent structural elements have to be connected. The connections can be hidden by decoration or cladding, in which case nobody is subsequently aware of their presence. They can be hastily detailed late in the design stage and left unclad, when they will clearly detract from any other attributes possessed by the building. Or they can be anticipated, painstakingly detailed and expressed in a truthful architectural manner.

A building in which all of the structural elements, that is the columns, walls, beams, floor slabs, roof slab and staircases are constructed of *in situ* reinforced concrete will not show any discontinuities between one element and the next. At the junction of a rectangular column and a rectangular beam, for example, there exists a rectangular solid of reinforced concrete which cannot be said to belong solely to one element or the other. It is part of both. Continuity is achieved by not curtailing either the vertical column reinforcement or the horizontal beam reinforcement. Both continue through the junction far enough to develop sufficient bond lengths with the concrete.

Connections between elements fabricated from structural steel can also be made so as to be almost imperceptible by the use of butt welding. This involves forming the weld over the entire section of the two parts of the element to be joined. If the welding is carried out efficiently, the joint has the same resistance to tensile, compressive and shear stresses as the connected parts. The integrity of butt welds carried out on site should always be checked by ultrasonic or gamma-ray methods. Whilst butt welds are useful for making site connections in members too long to be transported to the site in one length, the usual method of connecting a beam to a column is by welding angle cleats to the column, and bolting the beam through the flanges to these cleats. The welds in this case would be fillet welds, in which the weld metal is deposited around the perimeter of the cleats and the corresponding outline of the column. It is theoretically possible to connect any configuration of steel members together using a combination of welding and bolting.

The junction of two timber members in modern structures will always involve the intervention of a steel connecting device. Where the loads transferred across the joint are of a small magnitude, such as the shear forces that arise at the junction of a vertical stud and a sole plate, nails or screws may be adequate. Any connection between the principal members of a timber framework, however, will necessitate the use of bolts in conjunction with steel plates. In view of the greater likelihood of a framed structure in timber being left unclad than one of steel, owing to the greater fire resistance of timber, it is vital that the architect concerns himself at an early stage with the visual as well as the purely structural qualities of such a connection.

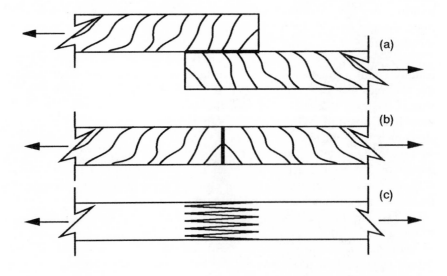

Figure 2.4   *Glue lines in timber*

Glued connections in timber are often thought of as equivalent to welding in steel, but this is an entirely false analogy. A glue line between two parts of a member in either tension or compression can only transmit that tensile or compressive force by acting in shear, as shown in Figure 2.4(a). Although the arrangement shown in 2.4(b) could temporarily locate the connected parts in a compressive joint, a glue line in direct tension has virtually no structural value. When it is necessary to connect two parts of a member whose axes lie in the same plane, a finger joint is used, as illustrated in Figure 2.4(c). As suggested by the name of this joint, the glue lines are formed at the interfaces of the fingers of wood projecting from each side. The alignment of the glue lines is almost parallel to the direction of the force, thus creating a structural behaviour more in keeping with that shown in Figure 2.4(a) than 2.4(b). The glue line is almost wholly in shear. The functioning of this joint can be simulated by interlacing the fingers of both hands, so that separation under an outward pull can only be prevented by tightening the fingers against each other, thus inducing a shear resistance between them.

Jointing of dissimilar materials

In a building where the reinforced concrete floors are carried across and supported on top of the masonry walls, the junction detail is relatively simple. Cross wall construction, as this system is known, involves building the walls up to the underside of the slab, casting the slab and building the next height of wall immediately above the wall below. The main danger here is in the possibility of setting out errors, resulting in misalignment of the upper and lower wall sections.

This particular connection between two different materials does not usually entail a great deal of creative thought at the design stage, mainly for two reasons. First, the floor load is applied more or less centrally within the wall thickness. Second, the floor relies on its own weight to keep it in position. Where, on the other hand, a horizontally spanning element is brought into the side of a vertical one, the matter is not quite so simple. The method of supporting a timber joist on a brick wall by means of a pressed steel joist hanger or a wall plate is a proven detail. Although usually concealed, the junction is deceptively neat because the joist does not encroach upon the domain of the wall.

Far less elegant and sometimes left exposed is the type of junction in which a steel lattice truss, for example, enters the side of a wall, whether of reinforced concrete or unreinforced masonry. The emergence of such details suggests that little thought has been given to the consequences of an arbitrary choice of materials for different structural elements. If an untidy detail, which may also be difficult to construct, can be anticipated, the structural system may need to be radically reassessed at the outset to facilitate both detailing and site operations. The steel trusses, for example, would probably be more easily connected to steel columns, thus changing the form of the building to that of a steel frame with masonry curtain walls. If a steel to masonry connection is unavoidable, the judicious use of corbelled supports could satisfy the demands of both function and aesthetics.

Architectural expression of joints

Where a deliberate change in material takes place, it is, if the junction is to be exposed, more conducive to an informed expression of structure to openly express rather than to conceal the transition from one material to another. This condition can only be successfully brought about if the geometry of both structural elements retains its integrity. The junction between the circular reinforced concrete columns and the laminated timber beams at The Burrell Collection Gallery in Glasgow (1983) shown in Plate 4 looks deceptively simple. But the steel plates emerging from the tops of the columns and bolted through the sides of the beams create a visual punctuation mark, an effect enhanced by the gap between the top of the column and the underside of the beam. The repetition of this simple detail adds to the architectural quality of the gallery space.

Another example of a simple, yet elegant, transition between two materials occurs in the exterior of The National Exhibition Centre in Birmingham (Plate 5), completed in 1980. The steel tie members are counterbalanced by the weight of the concrete bases, the junction between the two being made above ground level, and therefore permanently on view. Had this junction been detailed by taking the steel tie rods straight into the ground to an invisible concrete counterweight, the structural behaviour would have been less self-explanatory, and the steel ties more prone to corrosion. As detailed, the junction is a perfect expression of Newton's Third Law.

The aesthetics of structure in architecture is best served by freely acknowledging differences between materials rather than artificially concealing them.

Masonry junctions

It is to both the Classical and Gothic styles one must refer to understand the relationship between detailing and structural form. With stone as the definitive material in both of these traditions, tensile stresses had to be totally eliminated or kept to a very low level. It would be both misleading and trite to suggest that the column heads, such as that in Plate 1, by which a layman may recognize a classical building evolved solely as a device for punctuating a vertical and a horizontal structural element. Yet it would be well nigh impossible to effect a transition from entablature to

Plate 4 *Burrell Gallery, Glasgow*

Source: Teresa Lam

Plate 5   *National Exhibition Centre*

column, from a structural element in bending to one in compression, in a quarried material without the intercession of a capital. Even if the mason were able to cut the stones to perfection, a rectilinear entablature would rest awkwardly on a cylindrical column shaft.

The idea of a capital separating a column from the ribs of a vault is also typical of Gothic structural form, but there are cases where a smooth

Plate 6 *King's College Chapel*

transition from line to curve has been achieved. The interior of King's College Chapel in Cambridge (Plate 6), completed in the early sixteenth century, conveys the suggestion of an organic continuity of the fan vaulting to the shafts of the column. Given the ability of the masons to work the limestone into a curve, it would have been relatively simple to extend the lower portion of the stones at the springing of the fan vaults into a straight line. The identity of form between column and vault does nevertheless conceal a contrast in function. The vault is almost in a state of pure compression. The column, on the other hand, is being acted upon by a horizontal thrust as well as by a vertical load, so that it is constrained to behave as an eccentrically loaded, rather than as a concentrically loaded, column. The dimensions of the column need to be greater to support this combined load than would be the case if it were supporting only an axial compressive force. The required increase in size occurs on the outside in the form of a stepped pier. A masonry wall is not the most legible of structural elements from the interior.

Depending on the critical stance of the observer, the clean lines of the interior display either structural confusion, or an ingenious attempt at concealing the transition between two structural elements different in character.

Chapter 3

BEHAVIOUR OF BASIC STRUCTURAL ELEMENTS

The majority of built forms in architecture consist of rectilinear spaces. The greater part of most people's lives are spent in spaces bounded by horizontal and vertical planes. An appreciation of the behaviour of structural elements should therefore originate in the simple beam and column, both of which involve tension, compression and bending as described in Chapter 2.

Beams

Horizontal elements subject to bending

A structural element lying in a horizontal plane and subjected to vertical forces is commonly referred to as a beam. Using the vector notation introduced in Chapter 1, with downward pointing arrows representing gravitational forces and upward pointing arrows representing reactions, there are clearly numerous arrangements of loads and reactions that can be described as beams. A few of these are shown in Figure 3.1.

What all these arrangements, and countless others, have in common is that they all serve the purpose of moving vertical forces sideways to predetermined positions, that is to the supporting columns, walls and piers. The effect that these vertical forces have on the horizontal element, the beam, is to make it bend. The design, or the evaluation of the required size, of a beam must therefore be informed by the manner in which the bending effect caused by the external forces is resisted by the strength inherent in the beam.

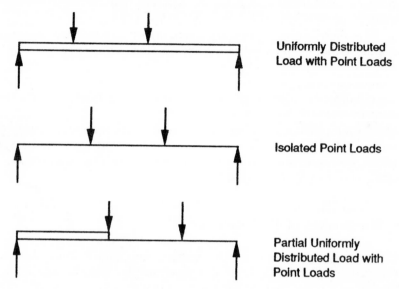

Uniformly Distributed
Load with Point Loads

Isolated Point Loads

Partial Uniformly
Distributed Load with
Point Loads

Figure 3.1 *Beam loading systems*

The effect produced by the external forces, including both loads and reactions, can be expressed as the bending moment, defined in Chapter 1. The term 'bending moment' was seen to be a function of the loading on a beam and its support positions. The more complicated the loading, the more lengthy are the calculations necessary to locate the position of maximum bending moment and to evaluate its magnitude. In most cases which occur in architecture, however, the design loading is based on assumptions which allow the use of the simpler load distribution patterns. The majority of beam design calculations can be carried out using the uniformly distributed load (UDL) or the central point load, for which the maximum bending moment can be quickly worked out from the formulae derived in Chapter 1.

It is not necessarily a wise decision to devise a structure where variations in the plan form between storeys create heavy point loads. This is sometimes unavoidable, but often leads to unacceptably deep beam sections. Structural design in architecture is as much about avoiding difficult situations as it is about solving lengthy mathematical problems.

Resistance moments

Recalling that a bending moment diagram is simply a graph, with the horizontal axis representing distance and the vertical axis the magnitude of bending moment, it can easily be inferred that the size of a simply supported beam is going to be governed by what is happening at mid-span. It is therefore necessary to possess a concept that will equate the maximum bending moment caused by the loading on the beam to the capacity of the beam to resist bending. This capacity is called the 'resistance moment'.

Several theories have been put forward over the last 400 years in an attempt to measure just how much a given cross-section of a particular material can resist a bending moment or, using the terminology just introduced, develop a resistance moment. However diverse the proposed solutions, all expressions for the value of a resistance moment involve the integration of two distinct variables:

- the strength in tension and compression of the material, and
- the geometry of the cross-section.

The relationship between tension, compression and bending has been introduced in Chapter 2. It is also essential in trying to visualize the behaviour of a beam that the idea of a 'couple' is fully understood.

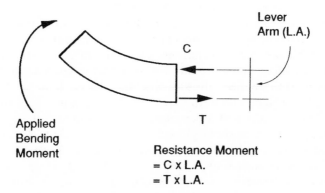

Figure 3.2 *Resistance moment couple*

Couples

A couple is a moment produced by two equal and opposite parallel forces. If these forces are thought of as the compressive and tensile forces, labelled C and T respectively in Figure 3.2, which are generated at any section of a bent beam, the relationship between bending and resistance moments becomes clearer. The value of the couple is the value of the moment produced by either one of these forces, and will, in keeping with the definition of the moment of a force, be equal to C or T multiplied by the perpendicular distance between them. This distance is known as the 'lever arm' when dealing with structural members, and is usually abbreviated to LA.

The resistance moment of the beam can therefore be expressed as

$$C \times \text{LA or}$$
$$T \times \text{LA}$$

This vector diagram expressing the relationship between externally applied and internally generated moments applies at any section along the length of the beam. Since, however, the greatest value of the resistance moment is required at the position of maximum bending moment, the magnitudes of C and T will also be at their maxima here. For a simply supported beam carrying a uniformly distributed load or a central point load, C and T will therefore have their greatest values at mid-span.

Elastic theory of beam design

All that now remains is the formulation of an expression for the resistance moment of a beam section in terms of those properties which contribute to the terms C, T and LA, in keeping with the known behaviour of the material concerned. It would be very convenient if all materials used in the formation of structural elements behaved according to principles of elasticity formulated by Robert Hooke, as discussed in Chapter 2. Steel certainly does exhibit a constant relationship between stress and strain up to its yield point. Broadly speaking, the more brittle the material, the more it will deviate from Hooke's Law. An assumption of elastic behaviour in timber, therefore, is more justified than a similar assumption for concrete. Codes of practice have, until quite recently, been drafted with Hooke's Law as the initial premise. More recently, however, they have tended to reflect the way in which loaded beams behave when tested to destruction. This principle, known as the 'ultimate load theory', seeks to establish the true factor of safety of a beam as the ratio of the load immediately before failure to the design, or working load.

The exploration of ultimate load design is beyond the scope and purpose of this book, and is best pursued through the medium of the current codes of practice. It is sufficient for the understanding of structural behaviour to examine a rectangular beam section on the assumption that the material obeys Hooke's Law within the range of the design loads on the beam. Instead of relating design loads to loads at failure, the safety factor is applied directly to the yield or failure stress of the material as appropriate. This figure varies considerably from one material to another, and is much smaller for steel, for example, made under conditions of strict quality control, than for timber, which has to be taken more or less as it comes from the tree.

The stress value thus arrived at is known as the permissible stress, and will be denoted throughout this book by the symbol f.

Rectangular sections

In Chapter 2 it was observed that the application of a moment to a test piece of an elastic material produces compressive stress and strain at one extreme fibre, and tensile stress and strain at the opposite extreme fibre. If the test piece is now thought of as part of a simply supported beam subjected to a uniformly distributed load, resulting in a parabolic bending moment diagram, the deflected form shown to some exaggeration would appear as in Figure 3.3, with the plane of zero stress, the neutral axis, at mid-depth.

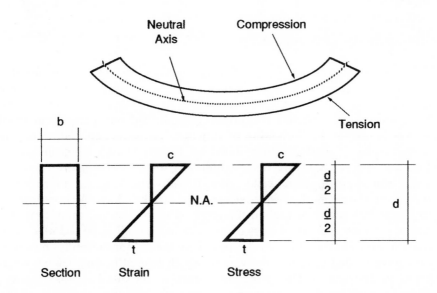

Figure 3.3 *Elastic bending of beam*

An assumption made in the elastic theory of beam design is that plane vertical sections through the beam before bending remain plane after bending. The section shown in Figure 3.3, in which the symbols b and d denote the breadth and depth of the beam respectively, is taken at the mid-span position, although the relationship between force, lever arm and resistance moment applies at any section. The strain diagram and the stress diagram drawn beside this section reflect the variation in strain and stress respectively throughout the depth d of the section. These diagrams are both graphs, in which the compressive strains and stresses are plotted to the right of the vertical axes, and the tensile strains and stresses are plotted to the left. This follows from the observed behaviour of a beam such as this in that the top half, that is above the neutral axis, is in compression, and the lower half is in tension. Since the strain diagram is linear, and stress is proportional to strain at every level in the depth of the beam, the stress diagram will also be linear.

Before calculating the value of the couple $C \times \mathrm{LA}$ or $T \times \mathrm{LA}$, which will be equal to each other, the importance of using the correct units must be emphasized. Since C and T are forces, they will be expressed in force units, in this case Newtons. The lever arm LA, depth d and breadth b, being dimensions, can be expressed in millimetres. The unit of the resistance moment is therefore Newton millimetres, abbreviated to N-mm. The hyphen in this case implies multiplication.

Figure 3.4 shows the relationship between the dimensions of the rectangular section and the stress diagram. The permissible stress for the material, which is critical at the extreme fibres, where the maximum horizontal ordinates on the stress diagram occur, is given the symbol f. This is measured in Newtons per square millimetre, abbreviated to N/mm².

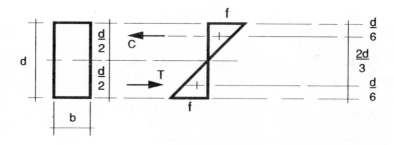

Figure 3.4   *Stress diagram through beam (elastic theory)*

The value of the force *C,* since force equals stress multiplied by area, will be equal to the average stress over the compressed area above the neutral axis times the area of the compressed half of the rectangular section, i.e.

$$C = \text{average stress} \times \text{area}$$
$$= f/2 \times bd/2$$
$$= fbd/4 \text{ Newtons}$$

Force *T,* relating to the stress and area below the neutral axis, will have the same value.

Lever arm

The value of the lever arm, the perpendicular distance between these two forces, depends on the location of the lines of action of *C* and *T* within the depth of the beam. These lines of action must coincide with the centres of gravity of the two identical triangular stress diagrams. In the case of a right-angled triangle, the centroid to which this corresponds occurs at a distance of one third of its height from the base. *C* and *T* are therefore located at one third of *d*/2 from the top and bottom fibres of the beam respectively.

In algebraic symbols, therefore,

$$C \text{ acts at } (1/3 \times d/2) = d/6 \text{ mm. from top}$$
$$T \text{ acts at } (1/3 \times d/2) = d/6 \text{ mm. from bottom}$$
$$\text{and LA} = d - (2 \times d/6) \text{ mm,}$$
$$= d - d/3 \text{ mm.}$$
$$= 2d/3 \text{ mm.}$$

Resistance moment calculation

With *C, T,* and LA derived in their appropriate units, the value of the resistance moment, abbreviated as *RM,* can now be expressed. Thus,

$$RM = C \times \text{LA (also} = T \times \text{LA)}$$
$$= fbd/4 \times 2d/3 \text{ N-mm}$$
$$= 2fbd^2/12 \text{ N-mm}$$
$$= fbd^2/6 \text{ N-mm}$$

This expression immediately reveals that the strength of a section in bending is proportional to d^2, that is to the square of its depth.

Also, the expression can be conveniently split up into two distinct parts, namely f and $bd^2/6$. Each part relates to one of the two variables identified earlier in this chapter, the first part to the strength of the material, and the second part to the geometry of the section. That part concerned with the geometry, $bd^2/6$, is known as the section modulus, and denoted by the shorter symbol z. This formula for the section modulus is only valid for a solid rectangular section. The z of any other section will have a different value, although in every case the depth d will appear as a squared quantity.

Rectangular timber sections

Timber joists

Having derived an expression for the resistance moment of a rectangular section, the most direct application is in the design of timber joists and beams. Sawn timber joists and laminated timber beams are nearly always rectangular in section. Once the maximum bending moment has been evaluated, and the permissible stress ascertained by reference to the relevant code of practice, the value of the depth d can be easily calculated, provided that the breadth b has been decided in advance. Thus, the expression equating resistance moment to bending moment, i.e.

$$RM = BM$$

can be expanded to

$$fbd^2/6 = BM$$

By multiplying each side by 6 and dividing each side by fb, this becomes

$$d^2 = \frac{BM \times 6}{fb}$$

Taking the square root of each side

$$d = \sqrt{\frac{BM \times 6}{fb}}$$

A simple example of a timber joist in a domestic floor will illustrate this application.

A timber floor, constructed of 50 mm wide timber joists spaced at 400 mm centres, is to span 4 metres. The superimposed, or live, load for domestic use is 1.5 kN/m². It is assumed that the average dead load of the joists, floor decking and ceiling has been estimated as 0.5 kN/m². The bending moment on each joist can be evaluated as follows:

$$\text{Dead load} = 0.5 \text{ kN/m}^2$$
$$\text{Live load} = 1.5 \text{ kN/m}^2$$
$$\text{Total floor loading} = 2.0 \text{ kN/m}^2$$

If the joists had been spaced at one metre centres, then the uniformly distributed load on each joist would have been 2.0 kN per metre run because each joist would have been supporting a loaded width of one metre, that is half a metre to each side of its centre line. Since the joists are spaced at 400 mm centres, however, the loading on each joist will be proportionally less.

$$\therefore \text{ Loading per joist} = 2.0 \times \frac{400 \text{ mm}}{1000 \text{ mm}}$$

$$= 0.8 \text{ kN per metre run.}$$

This loading is the uniformly distributed load (UDL) referred to in Chapter 1 by the symbol w.

Therefore the bending moment per joist will be equal to

$$wL^2/8 = \frac{0.8 \times 4^2}{8} = 1.6 \text{ kN-m.}$$

To express the resistance moment, a typical code of practice value for European softwood for f of 5 N/mm² will be used. The breadth b has already been fixed at 50 mm. Therefore

$$RM = fbd^2/6$$
$$= \frac{5 \times 50 \times d^2}{6} \text{ N-mm.}$$

Caution is necessary here in equating resistance to bending moments, in so far as the units on either side of the equation must be compatible. Since one expression has units of kilonewton-metres and the other has units of Newton-millimetres, it is convenient to convert the bending moment into the units of the resistance moment. Since there are 10^3 Newtons in a kilonewton and 10^3 mm in a metre, and $10^3 \times 10^3 = 10^6$, the expression for bending moment can be rewritten as:

$$BM = 1.6 \text{ kN-m} = 1.6 \times 10^6 \text{ N-mm}$$

$$\therefore \text{ since } RM = BM,$$

$$\frac{5 \times 50 \times d^2}{6} = 1.6 \times 10^6, \quad \text{from which}$$

$$d^2 = \frac{1.6 \times 10^6 \times 6}{5 \times 50} \quad \text{and}$$

$$d = \sqrt{\frac{1.6 \times 10^6 \times 6}{5 \times 50}} = 195.9 \text{ mm.}$$

This is the minimum theoretical depth which can be used. A smaller depth would, on examination of the formula for the resistance moment, increase the stress at the extreme fibres beyond its permissible level. Therefore the depth of joist required will be the next highest figure compatible with sawn sizes, which in this case will be 200 mm.

Deflections

In most cases where the size of a structural element is calculated as a response to an externally applied bending moment, additional checks have to be made. As spans become longer, deflection becomes more critical. Deflection is a measure of the maximum vertical displacement from the horizontal line joining the supports of the beam. Clearly, some limit has to be placed on deflections, large values of which can cause damage to ceilings, or produce an alarmingly curved appearance to the underside of a floor or supporting beam. Whereas the resistance of a beam to bending is proportional to the square of its depth, resistance to deflection is proportional to its cube. On the other hand, the mid-span deflection of a beam carrying a uniformly distributed load is proportional to the fourth power of its span, as compared with the variation in bending moment in proportion to the square of the span. The joist sizes quoted in tabulated form in reference books are therefore frequently governed by deflection, and may show depths greater than those dictated solely from consideration of resistance moments.

Holes through joists

The stress distribution through the depth of the section reveals that the timber is only working at its full capacity at the extreme fibres. The compressive and tensile stresses both decrease towards the interior of the section, eventually becoming zero at the neutral axis. Awareness of this condition is useful if quick site decisions need to be taken, in that small holes for electrical conduits and heating pipes can be accommodated near the centre of a joist without great detriment to the strength of a section. It is never advisable to cut holes near to the top and bottom of a joist, where the bending stresses are greatest, particularly in the vicinity of the position of maximum bending moment. Sudden changes in section should always be avoided where possible, no matter what the material or form, and permitted only after close investigation of the strength of the reduced section.

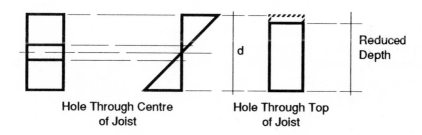

| Hole Through Centre | Hole Through Top |
| of Joist | of Joist |

Figure 3.5   *Holes through joists*

Figure 3.5 shows that the small hole with its centre at the neutral axis removes the least critically stressed timber, at the same time preserving the original depth of the joist. The hole at the top of the joist removes critically stressed timber, also leaving a section with a reduced depth.

Whereas bending stresses are highest at the extreme fibres, the shear stress at any section, in contrast, reaches its maximum value at the neutral axis, and is equal to zero at the top and bottom of the section. Since the shear stress is directly proportional to the shear force at any section, it is as well to avoid cutting any holes through the centre of a joist near to its supports. For this reason, pipes and conduits through the centre of a joisted floor should be positioned nearer to mid-span than to the supports.

Joists laid sideways

It was not untypical in timber buildings erected during the Middle Ages to lay rectangular joists on their sides, with their smaller side vertical. This may seem an absurd thing to do in the light of what is now known about structural behaviour. It must have seemed, however, quite reasonable at the time to arrange the sawn timbers forming a floor in the position in which they were least likely to fall over during construction. A single joist laid in the position now known to be correct, with the greater dimension vertical, is easier to dislodge than one laid on its side.

A comparison between the section moduli of a particular sawn size will demonstrate the order of the difference in the resistance moments for these two conditions. The z value of a section 150 mm $\times$ 50 mm in the correct position is

$$bd^2/6 = 50 \times (150)^2/6 = 187500 \text{ mm}^3$$

whilst the z value with the joist on its side is

$$db^2/6 = 150 \times (50)^2/6 = 62500 \text{ mm}^3$$

The reduction in strength will always be in proportion to the ratio of the side lengths b and d, which in this case is one-third.

It can also be shown that the difference in deflection at mid-span will vary as the square of the ratio of d to b. The deflection of the 150 mm $\times$ 50 mm section would therefore be nine times greater under the same load when positioned incorrectly.

Effect of doubling breadth

The manipulation of the values of b and d in the formula for the section modulus can also be used to demonstrate that doubling the breadth of a joist does not mean the depth can be halved. Given that $z = bd^2/6$ and that it is decided to reduce the joist depth by doubling the value of b to $2b$, the required depth can be called d_1. For the two z values to be identical, therefore, they can be algebraically equated, i.e.

$$z = 2b(d_1)^2/6 = bd^2/6$$

Dividing each side by $b/6$,

$$2(d_1)^2 = d^2$$
$$\therefore (d_1)^2 = d^2/2$$
$$\therefore d_1 = d/\sqrt{2} = 0.707d$$

When two joists of identical breadth are used side by side, therefore, their depths can be reduced to seven-tenths of that of the originally proposed single joist, but only when the design is governed by resistance to bending. It can be shown that in order to keep the deflection to the same limit as that of the single joist, the depth of the double joist needs to be eight-tenths of the depth of the single joist.

Laminated timber

The maximum spans for sawn timber joists are mainly governed by the lengths and section sizes that can be obtained from the felled tree and transported in large quantities. If deeper sections are needed for longer spans, built-up members can be manufactured in the form of laminated, box, or 'I' section beams.

Box and 'I' sections involve the use of plywood or hardwood webs, to which softwood compressive and tensile flanges are attached by glueing and nailing. The reasons governing the choice of these forms are analogous to those underlying the fabrication of similarly shaped sections in steel, discussed in Chapter 4. The design procedure for a laminated timber beam is similar to that for a sawn joist.

A laminated timber beam is formed by sawing the original timber into thin strips, and glueing them together to create a solid section. The width of the individual strips, or laminations, and therefore the breadth of the beam, is usually at least 100 mm, the normal thickness being 50 mm. The length of the laminations does not limit the length of the beam, since adjacent lengths can be connected by finger joints, of which there should ideally be only one in any vertical section of the beam. There is no theoretical limit to the length of such a beam. There is, however, a point at which it becomes wasteful to meet the demands of increasing spans by providing deeper solid sections, no matter what the material. Practical size limitations are also imposed by considerations of storage, transportation and erection.

The glue lines between the laminations are formed under carefully controlled conditions to ensure a high resistance to horizontal shear, that is the sliding of one lamination on the next one above or below. If this were to happen, the system would consist not of one vertically continuous section but of a succession of individual laminations, with each one passing its load on to the one below. In the extreme condition of totally frictionless interfaces between all of the laminations, the bottom lamination would be trying to act as a 50 mm deep beam, obviously a hopeless task.

Reinforced concrete

Structural behaviour

In Chapter 2, a distinction was drawn between those materials from which structural elements are formed during the construction process, and those which lend themselves to preparation or fabrication before work on site begins. Timber is a naturally occuring material which can be sawn to the dimensions needed for posts, joists, purlins, rafters and other elements of relatively small cross-sectional areas and lengths. It can also be built up into the laminated form described in the previous section. Structural steel is manufactured under conditions of strict quality control, and is also admirably suited to the production of linear elements such as beams and columns, or stanchions as they are frequently referred to in this material. Since steel has strength properties greatly in excess of those of timber, the limit of potential spans and loads is clearly much greater.

Reinforced concrete structures are, on the other hand, conditioned by the manner in which they are produced. Provided that the formwork can be made, any shape can be created in reinforced concrete. Although the use of this material, in the majority of buildings, results in prismatic members such as beams and columns of rectangular section, it is as well to remember that concrete possesses an innate sculptural quality. Provided that the laws of structural mechanics in the earth's gravitational field are acknowledged, there is no limit to the structural forms that can be created.

The behaviour of reinforced concrete as a rectangular element subjected to a bending moment allows some comparisons to be drawn with a rectangular timber section. The obvious difference is that timber, being composed of one substance, is homogeneous whilst reinforced concrete, being composed of two materials, is heterogeneous.

Whereas the whole section of a timber beam is available for generating a resistance moment, this is not the case with reinforced concrete. In a simply supported beam, there will be compression above the neutral axis and tension below. Since concrete is strong in compression and weak in tension, only the concrete above the neutral axis will be structurally effective. Below the neutral axis, the concrete in tension will be cracked, and will therefore not contribute a force towards the resistance moment couple. The longitudinal steel reinforcement is this zone supplies the tensile force T to balance the compressive force C which together make up the two equal and opposite forces of the couple.

Rectangular reinforced concrete beams

For a long period following the introduction of reinforced concrete into the design of building structures, design theories imitated those described earlier for timber sections. It was assumed that within a certain range, a concrete test specimen would behave in accordance with Hooke's Law, with stress proportional to strain. The stress diagram headed 'Elastic Theory' in Figure 3.6 emerged, in which the compressive stress in the concrete varied as a straight line from a maximum at the extreme fibre, in this case the top surface, to zero at the neutral axis. Below the neutral axis, the concrete is subjected to tensile deformation, tensile strain and tensile stress, and is therefore ignored. The zero stress value is maintained until the centre of gravity of the group of reinforcing steel bars is reached. At this position, the tensile force T can be expressed as a vector without actually drawing a stress diagram. T will still be evaluated as a stress multiplied by an area, i.e.

$$T \qquad = \qquad A_{st} \qquad \times \qquad P_{st}$$

| (Force in Newtons) | (Area of steel in mm²) | (Permissible steel stress in N/mm²) |

but the diagram is simplified by simply showing T on its line of action. The lever arm LA will be the vertical distance between the lines of action of C and T.

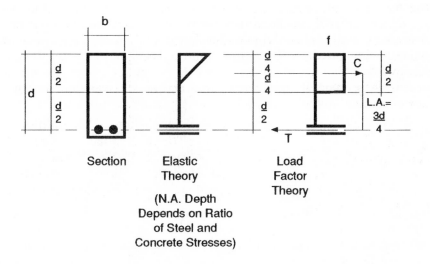

Figure 3.6 *Reinforced concrete section in bending*

As research continued into the way in which reinforced concrete beams respond to load, it became apparent that as a beam approached failure, quite dramatic redistributions of stress occurred. The fibres near to the neutral axis would gradually accept more compression until, just before the time of failure, the stress distribution above the neutral axis would resemble a rectangle instead of a triangle. The verb 'resemble' is carefully chosen, since the highest stress value in the concrete tends to occur not at the top, or extreme fibre in compression, but a little way down. Moreover, the precise shape of the stress diagram varies according to the strength of the concrete from which the beam is cast. However, the approximately rectangular shape of the stress diagram gave rise to a convenient method of using an elastic, or working load, design theory, whilst taking advantage of the hidden potential for higher compressive stresses close to the neutral axis.

This stress distribution is shown in the last diagram in Figure 3.6, and can be used to give a more reliable estimate of the resistance moment. A value of the permissible stress f is arrived at by dividing the cube or cylinder crushing strength of the concrete by a series of safety factors. If the assumption is made that this stress can be relied upon throughout the entire top half of the beam, the depth of concrete in compression will be $d/2$, where d is the 'effective' depth of the beam. The concrete below the steel, known as the cover, can be ignored for calculation purposes. It is there principally to protect the steel against corrosion, and from the action of fire. It may be remembered from Chapter 2 that unprotected steel has very little innate fire resistance. The cover is also vital to the provision of an adequate bond between the steel and the concrete. If this were insufficient, the steel could slide horizontally within the sleeve of concrete, with consequences similar to those arising from the failure of a glue line in a laminated timber beam. Both are related to horizontal shear stresses which, if not acknowledged and resisted, would destroy the vertical continuity of the section.

The algebraic value of C is therefore f multiplied by the area subjected to constant stress, i.e. $C = fbd/2$ Newtons. The value of the lever arm, LA, can be found by first identifying the position of the line of action of C. This is easier to locate for a rectangular stress block than for a triangular one, since the centre of area of the rectangle is half way down, that is at $d/4$ mm from the top.

$$\therefore \text{LA} = (d - d/4) = 3d/4 \text{ mm}$$

The resistance moment, in terms of the compressive strength of the concrete, can therefore be evaluated as follows:

$$RM = C \times \text{LA}$$
$$= fbd/2 \times 3d/4 \text{ N-mm}$$
$$= 3fbd^2/8 \text{ N-mm}$$

The compressive stress of the concrete is influenced partly by the amount of cement used with the aggregates. It is possible to specify concrete in terms of the proportions of cement to fine to coarse aggregate (e.g. 1:2:4) or to aim for a particular level of cube or cylinder strength. The value of f used to express the resistance moment of 1:2:4 concrete would be about 4.7 N/mm². The crushing strength of a 28-day-old cube taken from this mix would need to be at least 21 N/mm². The potential resistance moment for a rectangular reinforced concrete beam of breadth b mm and depth d mm, using a 1:2:4 concrete mix, would be:

$$\frac{3fbd^2}{8} \quad \text{N-mm}$$

$$= \frac{3 \times 4.7 \times bd^2}{8} \quad \text{N-mm}$$

$$= \quad 1.76\ bd^2 \quad \text{N-mm}$$

The potential resistance moment of a beam using a 1:1½:3 mix, for which the 28-day mimimum cube strength is 25.5 N/mm² and f is 5.7 N/mm², can be worked out in a similar manner, or as a proportion of the relative cube strengths. Thus,

$$RM \text{ of } 1:1\frac{1}{2}:3 \text{ rectangular section} \quad = 1.76\ bd^2 \times \frac{25.5}{21} \quad \text{N-mm}$$

$$= 2.13\ bd^2 \quad \text{N-mm}$$

$$\text{Similarly, } RM \text{ of } 1:1:2 \text{ rectangular section} \quad = 1.76\ bd^2 \times \frac{30}{21.5} \quad \text{N-mm}$$

$$= 2.5\ bd^2 \quad \text{N-mm}$$

The adjective 'potential' has been used to describe these resistance moments because the true resistance moment of the section is going to be conditioned by the amount of longitudinal reinforcement provided. If the applied bending moment is as high as the resistance moment generated by the compressive strength of the concrete, then the tensile force T will have to be of the same magnitude as C. The cross-sectional area of reinforcement can then be found by dividing C by the permissible stress in the steel. This relationship is clarified if it is recalled that since

$$\text{Force} \ = \text{Stress} \times \text{Area}$$

$$\text{Area} \ = \ \frac{\text{Force}}{\text{Stress}}$$

Since $C = T$, therefore,

$$A_{st} \ = \ \frac{C\ (=T)}{P_{st}} \quad \text{mm}^2$$

If the applied bending moment is smaller than the potential resistance moment of the section, the reduced value of T can be calculated by dividing the bending moment by the lever arm, and substituted for C in the above expression.

Reinforced concrete 'Tee' sections

Having described reinforced concrete almost as a derivative of timber in focusing attention on a rectangular section, it may seem frivolous to add that this shape is not the most efficient, nor the most widely employed in structures formed from this material. In fact an isolated rectangular section is quite rare. It can appear in the form of a lintel carrying brickwork or blockwork over a wall opening. It occurs sometimes as a beam supporting floor or roof elements constructed of a different material, for example steel or timber purlins. Whilst the majority of horizontally spanning elements appear to be rectangular when looked at from underneath, closer inspection usually reveals that they have been cast integrally with the floor slab above them. Whether the system is spanning in one direction, for example as a beam cast at the same time as the supported slab, or in two directions as the ribs and flange of a waffle slab, the effective section ceases to be rectangular. The structural strength of the section becomes that of both the horizontal slab and the downstand rib. The resulting section is known as a 'Tee' section, as shown in Figure 3.7.

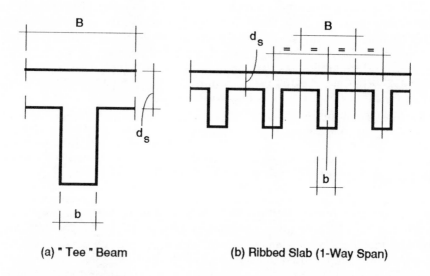

(a) " Tee " Beam   (b) Ribbed Slab (1-Way Span)

Figure 3.7   *'Tee' sections*

In both sections shown, it is impossible for the vertical rib of the 'Tee' section to deform without taking with it the immediately adjacent parts of the slab. This action brings about horizontal shear stresses along the planes of potential separation. These stresses are normally within the capacity of the concrete. Even when they are not, these planes will be made less susceptible to shear failure by the presence of vertical stirrups or horizontal slab reinforcement. The 'Tee' section can thus be considered as a single structural element, in which the breadth of the compressive zone is greater than the breadth *b*. The increased breadth *B* varies according to parameters such as the span of the beam, the rib breadth *b*, the spacing of the ribs and the slab thickness d_s. In Figure 3.7(a), *B* will depend mainly on rib breadth and slab depth. In Figure 3.7(b), where the ribs are closely spaced, *B* will clearly be equal to the spacing of the ribs. Under such conditions, the area of concrete available for generating the compressive force *C* of the resistance moment couple is greatly in excess of the area above the neutral axis of a rectangular section. Hence the required depth of a reinforced concrete 'Tee' section is rarely governed by compression. On the contrary, what often happens is that the depth *d* has to be increased from the assumed or calculated value in order to accommodate sufficient steel reinforcing bars to provide the tensile force *T*. There is a limit to the number of bars that can be squeezed into the comparatively narrow rib of a 'Tee' beam without causing difficulties in pouring the concrete between them. By making the section deeper, the lever arm increases, resulting in a decrease in the magnitude of the tensile force *T* and in the amount of steel required.

There is one condition in which a reinforced concrete beam reverts to its unnatural rectangular state from the geometrically more favourable 'Tee' shape. This occurs when the positions of the lines of action of the compressive force *C* and the tensile force *T* are reversed, as for example around the supports of a continuous beam or throughout the length of a cantilever. In these circumstances, the slab adjacent to the top of the beam is in the tensile rather than the compressive zone, and does not therefore create any increase in the resistance moment. Cantilevered and continuous beams are covered in Chapter 7.

Columns

As beams are horizontal members whose dimensions are essentially determined by the bending moments imposed on them, columns are vertical members whose lateral dimensions are primarily conditioned by the load they are required to carry, and by their height.

Our perception of the apparent strength of a column in a built form is to a certain extent a visual one. An isolated column, one metre or more square, in the ground storey entrance hall of a multi-storey building, may look appropriate to the imagined load bearing down upon it. Equally subjective is the experience of contemplating a series of closely spaced columns, 150 mm square, in a single storey building, or as a colonnade, in which context they may not look alarmingly fragile.

The classical orders

If an inhabitant of Western Europe were asked to sketch what he understood by a column, a form not unlike one of the classical orders would probably appear. An order in classical architecture is generally taken to mean the unit of a Greek temple colonnade, consisting of a column and its superstructure. The four best known orders – the Tuscan, Doric, Ionic and Corinthian (Plate 1) – were first described by the Roman architect Vitruvius. A fifth order, the Composite, appeared in the writings of the Renaissance architect Alberti. Each of the columns in the five orders have their own rules on slenderness and proportion. Each column, moreover, has its own distinctive column head, from the simple square and circular geometry of the Tuscan to the elaborately carved acanthus leaf of the Corinthian and Composite. The adoption of one order rather than another for a particular building in Greek, Roman and Renaissance architecture was not the result of an arbitrary choice. It was an acknowledgement of the way in which each order could express its own meaning in terms of the function and purpose of the building.

Although the earliest temples were constructed of wood, the five orders are usually historically associated with some form of stone. The reason for the original presence of a column head, leaving aside the symbolism inherent in the five orders, could well have been based on considerations of a more practical nature. A column terminating beneath the entablature would have presented an untidy junction. It would probably have been more difficult to ensure that the load was supported near to the centre of the column – a necessary condition if tensile stresses were to be avoided.

Functions of the column

To describe a column in terms of its most obvious function, one need go no further than to state that it is there simply to transfer down to the base a load applied at its head. All columns, whether they belong to the primitive, Classical, Gothic, modern or any other architectural style, whether they be beautiful or ugly, have to justify their existence in this way. Yet a much clearer impression of their structural behaviour emerges when a column is regarded as a particular form of a strut – the name given to a compression member. A column is a vertical strut. A tie, being subject to a tensile force, will remain straight and in a state of tension until failure is reached. A strut, unless very short in relation to its cross-sectional area, will not remain in its original straight condition as it begins to fail. It will be noticeably bent, implying that it is subject to bending moments throughout its height as well as to axial compression. The bending moments so caused will, with increasing load, eventually lead to failure of the column by buckling. Bending can only be absent from a column, even under working loads, in the purely theoretical condition when the applied load does not waver from the axis in the journey from head to base.

Columns in the elevations of a building are also called upon to transmit wind loads as horizontal uniformly distributed loads to the floors, which then act as horizontal reactions. This creates bending moments in the columns as though they were beams rotated through 90°. Where the construction is monolithic, as with reinforced concrete, it is impossible to avoid the transmission of bending moments at beam to column junctions. The same applies to some types of connection in structural steelwork and timber. It is essential, therefore, when looking at or thinking about any column built from any material, to understand that its function as a structural member in an apparent state of pure compression is dependent to a certain extent on its resistance to bending.

Hollow steel columns

Steel being the strongest available structural material, it is not surprising that if heavy loads are to be carried by the smallest amount of material, a steel section is the obvious choice. For slender columns, that is columns which are tall in relation to their cross-sectional area, the buckling load can be used as a basis for estimating the loads which they can safely carry. For less slender columns, more squat in appearance, the buckling load will give a misleading impression, erring on the high side. Other methods which bring into play the axial stress on the column section are more reliable here.

The buckling load on a column is inversely proportional to the product *EI*, where *E* is the modulus of elasticity defined in Chapter 2. *I* is a function of the geometry of the cross section, and is the algebraic symbol for the second moment of area. *I*, for a given sectional area, will increase with the square of the distance by which the material is remote from the centre of the area, and is measured in mm^4. This can be compared with the section modulus *z*, encountered earlier in this chapter as an index of resistance to bending, which is measured in mm^3 and also known as the first moment of area. If one were searching for the most efficient shape for a column, therefore, the circular or square hollow section would emerge as the most likely possibilities. Even without ever having heard of second moments of area, observation of analogous forms in nature, such as the stems of plants, would lead to this conclusion. The resistance to buckling of a drinking straw is a man-made illustration of this principle. The thin membrane of the straw, if flattened out, would resist only a very small compressive load before buckling, as well as being quite useless as an aid to the upward movement of thirst-quenching liquid.

The width of the material at the perimeter of a hollow section is called the 'wall thickness'. The further this wall is situated from the centre of the section, therefore, the greater is the *I* value, and therefore the theoretical buckling load. There is, however, a limit to this lateral expansion of an area, in that if the process is taken to extremes, there would emerge a tube with a second moment of area approaching infinity and a wall thickness approaching zero.

Returning to the real world, circular hollow sections are manufactured by the British Steel Corporation with external diameters between 21 and 457 mm, with wall thicknesses varying from 3.2 to 40 mm. Every diameter has a range of wall thicknesses, so that the external measurement of a steel tube is not a precise indication of its load-bearing capacity. For example, a 457 mm diameter tube of grade 43 steel will, over an effective height of three metres, safely support a load of 7789 kN if the wall thickness is 40 mm, and 2087 kN if the wall thickness is 10 mm. These loads, incidentally, would not be determined solely on the basis of the buckling load, since the columns here are relatively squat, with a visual height to diameter ratio of less than 7.

For a constant area of section and wall thickness, the circular shape is the most efficient hollow section for axially loaded columns. A rectangular hollow section so used is less efficient because a strut will always buckle about the weaker of the two principal orthogonal axes. Rectangular hollow section columns are most effective where bending moments induced by fixity, as in a portal frame, are more critical than axial forces. Fixity and portal frame action are discussed in Chapter 5.

Steel 'I' section columns

A steel 'I' section beam is so formed as to deploy most of the available material where it is most needed, that is at the extreme fibres within the plane of bending (see Chapter 4). An 'I' section column, on the other hand, needs to have more of a square enclosing outline to provide first and second moments of area of a similar order about the two principal axes. This section is known as the 'universal column', often abbreviated to UC. As a comparison, a 254 mm × 254 mm × 132 kg/metre UC and a 350 mm × 350 mm square hollow section have identical sectional areas. Over a three-metre effective height, the UC can safely carry 2286 kN as an axial load, whereas the hollow section will support the higher load of 2490 kN. The 'square I' section is slightly weaker than the hollow section mainly because some of the material, that is the web, is aligned with one of the neutral axes. An advantage in using a universal column is that as with an 'I' section beam, it is comparatively easy to connect other components through the flanges and web with bolts. Connections to hollow sections away from the top and the base always involve welding plates on to the outside face. This is, incidentally, easier with a square than a circular section, whose perfect Platonic shape it seems a pity to violate.

Reinforced concrete columns

The principal role of steel reinforcement in beams is to provide the tension force in the resistance moment couple, thus compensating for the lack of tensile resistance in the concrete. In reinforced concrete columns, the main function of the longitudinal steel bars is to increase the load-carrying capacity possessed by the concrete alone. To avoid any oversimplification, there are cases when the reinforcing steel is also placed in the compression zone of a beam in order to increase the resistance moment over that of the concrete alone. Column reinforcement may also be in tension in the presence of load eccentricities and applied bending moments.

As well as adding to the compressive resistance of a cross-sectional area of concrete, the reinforcing bars, being located as near as possible to the faces of the column and tied at frequent vertical intervals with steel transverse links, are well placed to arrest any incipient tendency to buckle. For this reason, reinforced concrete columns can be made much more slender than piers of unreinforced masonry.

In common with hollow steel columns, it is not possible to judge the strength of an existing reinforced concrete column solely from appearances. The total area of steel can vary between 0.8 and 8 per cent of the gross area of the concrete section, although 4 per cent is considered to be a practical maximum value to allow for overlapping of bars. Taking as an example a 300 mm × 300 mm section of a column using a 1:2:4 concrete mix, its load-carrying capacity can be evaluated by adding together the separate capacities of the concrete and of the steel. The permissible stress in direct compression for this mix of concrete (p_c) is 5.3 N/mm² and for the mild steel bar reinforcement (p_{sc}) it is 125 N/mm². The gross area of the section is (300 × 300) = 90,000 mm²

$\therefore$ with 0.8% reinforcement, steel area $(A_{sc}) = (0.8\% \times 90,000) = \quad 720$ mm²
and concrete area $(A_c) = \quad (90,000-720) \quad = 89,280$ mm²

$\therefore$ permissible load on steel $= A_{sc} \cdot p_{sc} = (720 \times 125 \times 10^{-3}) \quad = \quad 90kN$
and permissible load on concrete $= A_c \cdot p_c = (89,280 \times 5.3 \times 10^{-3}) = 473kN$
$\therefore$ capacity of column $\qquad\qquad\qquad\qquad = 563kN$

With 4% reinforcement, steel area $(A_{sc}) = (4\% \times 90,000) \quad = \quad 3,600$ mm²
and concrete area $(A_c) = (90,000-3,600) \quad = 86,400$ mm²

$\therefore$ permissible load on steel $= A_{sc} \cdot p_{sc} = (3,600 \times 125 \times 10^{-3}) = 450kN$
and permissible load on concrete $= A_c \cdot p_c = (86,400 \times 5.3 \times 10^{-3}) = 458kN$
$\therefore$ capacity of column $\qquad\qquad\qquad\qquad = 908kN$

The figure 10^{-3} converts the loads on the steel and the concrete from newton into kilonewton units.

The load-carrying capacity of the column is therefore increased by a factor of $908/563 = 1.61$ when the proportion of steel is increased from .8% to 4%.

These permissible loads are valid when the slenderness ratio of the column, that is the ratio of effective height to the least lateral dimension, does not exceed 15. Beyond this figure, a reduction factor has to be applied, resulting in decreasing permissible loads for progressively more slender columns. For a slenderness ratio of about 30, the load is halved. At a slenderness ratio of 57, one code of practice stipulates that the column can carry no load at all, not even its own weight.

As pointed out earlier in connection with reinforced concrete beams, the current code of practice adopts the procedure of relating failure loads to ultimate stresses, instead of working loads to permissible stresses.

As the percentage area of reinforcement is increased, therefore, the column is attempting to transform itself from a concrete structural member into one of steel, this condition being reached when the steel area is 100 per cent – an impossible state of affairs, but a reminder that a concrete column in a completed building does not reveal the source of its inner strength.

This variation in the strength of a column according to the quantity of reinforcement is a useful characteristic when the architect wishes to keep all column dimensions the same for the total height of a building. The response to increasing load is to provide correspondingly greater reinforcement areas.

Slender columns in architecture

Architects and students are often pleasantly surprised to discover that reinforced concrete columns do not have to be so large as they had anticipated. The notion of a column size in relation to its height is usually informed by visual reference to the classical orders, and therefore to unreinforced masonry. Furthermore, most columns in buildings of reinforced concrete tend to be oversized owing to considerations such as fire resistance and the ease of pouring the concrete between the steel bars. Tall slender columns in reinforced concrete are rare, but when introduced as a structural element in an architecturally important space (is there such a thing as an architecturally unimportant space?) can make a very bold statement, simply because small lateral dimensions are unexpected. Plate 7 shows one example of such a phenomenon. The foyer of Denys Lasdun's National Theatre in London (1976) has rectangular columns where the slenderness ratio, that is height to breadth, is about 30. Notre Dame Le Raincy (1924), near Paris, designed and built by Auguste Perret, has circular columns in the nave whose slenderness ratio, in this case height to diameter, is of a similar order. The use of a rectangular section in the National Theatre enhances the sense of drama in that the direction of potential buckling, about the weakest axis, is visually defined.

Plate 7 *National Theatre*

Most reinforced concrete columns in modern architecture have slenderness ratios of 15 or less, which classifies them in code of practice terms as short columns. For this reason, they tend not to be noticed, particularly when constructed in the plane of the external enclosures. An architecture rich in the expression of reinforced concrete columns, with a few notable exceptions, has yet to emerge. Much could be learned by the rediscovery of the language inherent in the finest classical buildings, in which proportion and material are compatible.

Unreinforced masonry

The absence of tensile strength and the interstratification of mortar joints, usually weaker than the brick or block, leads to smaller permissible loads in piers of brickwork or blockwork than for reinforced concrete columns of similar dimensions. BS 5628: Part 1 stipulates a reduction in the basic design compression stress for slenderness ratios in excess of 8, as compared with 15 in reinforced concrete. The maximum slenderness ratio quoted is 27, at which the reduction factor for an axial load is 0.40. As a comparison, consider a 300 mm $\times$ 300 mm brick pier in which the bricks have a crushing strength of 20.5 N/mm², almost identical to that of the 1:2:4 concrete mix used earlier as the basis of the calculations for reinforced concrete members. The basic design stress given in the code, using a 1:1:6 (cement:lime:sand) mortar mix, is 1.3 N/mm². Using an effective height of 3 metres, the slenderness ratio is

$$\frac{3 \times 10^3}{300} = 10$$

for which the code specifies a reduction factor of 0.97 for axial loads.

$\therefore$ permissible stress on column, or pier, $= (1.3 \times 0.97)$ $= 1.26$ N/mm²
$\therefore$ permissible load on column $= (300 \times 300 \times 1.26 \times 10^{-3}) = 113.4$ kN.

This is only one-fifth of the load obtained for the 1:2:4 concrete column reinforced with 0.8 per cent longitudinal steel.

Another factor which reduces the strength of this pier is the comparatively low strength of the mortar, for which the mean compressive strength at 28 days after mixing is only 2.5 N/mm².

It is possible to enhance the capacity of a brick or block pier by using perforated units through which longitudinal bars can be threaded, thus endowing the member with more resistance to buckling. The load-bearing capacity will still, however, be limited by the lower mortar strength. This device is nevertheless extremely useful when it is desired to create large openings, therefore inducing heavy point loads rather than lighter distributed loads, in a load-bearing masonry structure. To change the material and introduce random steel or reinforced concrete columns in a masonry structure may involve complications and delays in construction, not to mention a visual intrusion.

Timber columns

Because of its ability to resist tensile as well as compressive stresses, it is intuitively felt that a timber column, or post, would be more effective in resisting the inevitable buckling tendency than one of similar dimensions in unreinforced masonry. This is reflected in the provisions of the current code of practice for timber, BS 5268, in which slenderness ratios of up to 46 are allowed, as compared with 27 for piers and walls in unreinforced masonry. In view of the wide variation in the basic permissible stresses for timber, comparisons with other materials do not lend themselves to a definitive presentation. It is nevertheless of some interest to examine the permissible axial long-term load, that is the dead plus permanent imposed loads, for the range of softwoods available. These are divided in BS 5268 into five strength classes, each of which may contain several species of timber. By the same token, the same species may appear in three or more strength classes. The basic compressive stresses parallel to the grain for the lowest and the highest of these strength classes, referred to as SC1 and SC5 respectively, are 3.5 N/mm^2 and 8.7 N/mm^2.

Using the same effective height (3 metres) and cross-sectional area (300 mm × 300 mm) as in the previous section, the reduction factor for the slenderness ratio of 10, taken from the code of practice, is 0.92. The permissible loads are:

$$SC1 \quad (0.92 \times 3.5 \times 300 \times 300 \times 10^{-3}) = 289.8 \text{ kN}$$
$$SC5 \quad (0.92 \times 8.7 \times 300 \times 300 \times 10^{-3}) = 720.4 \text{ kN}$$

Even for the lowest strength class, the load carrying capacity is far greater than the 113.4 kN permissible load on the unreinforced masonry pier in the previous section. The load for the timber column in strength class 5 is, moreover, greater than that for the reinforced concrete column with the mimimum amount of reinforcement calculated as 563 kN.

It can also be shown that the 113.4 kN load on the brick or block pier could equally well be carried by a timber section 150 mm × 150 mm from strength class 1, with a lot in reserve, i.e.

$$\text{Slenderness ratio} = \frac{3000}{150} = 20, \text{ for which the code stipulates a reduction factor of 0.77.}$$

The permissible load is therefore $(0.77 \times 8.7 \times 150 \times 150 \times 10^{-3}) = 150.7$ kN.

It is possible to make endless further comparisons between the strongest timber and the weakest masonry, the weakest timber and the strongest masonry, least available timbers with most easily obtainable bricks, without being able to make any profound value judgements about the relative merits of these two classes of material. The choice of one or the other should normally be determined by the nature of the whole building structure rather than by a more local consideration. There may be cases in rehabilitation work, however, when a load from above has to be carried using the smallest sectional area. In terms of simple compressive strength, steel may well be a better solution in such circumstances, but the need to provide fire resistance could well make timber the optimum choice, since the steel surface would need protecting.

The permissible loads for laminated timber columns will again vary with the strength and species of timber used, but will have a higher upper limit than for sawn softwoods. Whereas the option exists for using laminations of a lower strength near to the centre of a laminated timber beam, that is close to the neutral axis, a laminated timber column supporting an axial load will be stressed evenly over the whole section, and will not permit such variations in grade.

Chapter 4

BEAM AND TRUSS SYSTEMS

A truss appears to be a structural element in which small components are assembled into a larger form compatible with the distance to be spanned. Contemplation of the assembly process at once suggests a structural form totally different in kind from a beam. Nothing could be farther from the truth. The truss and the beam are analogous forms. Both are systems which, when used as horizontal members, displace vertical forces to positions of vertical reaction, thus attracting bending moments. Depending on which mental image is preferred, the truss can be thought of as a hollow beam, or the beam as a solid truss.

Growth of solid beams

The fundamental structural logic underlying the adoption of a trussed form can be appreciated by considering what happens to the stress at the extreme fibres of a rectangular beam as it doubles its size. Figure 4.1 shows two such beams, one with depth d, breadth b, and span L, and the other with dimensions of $2d$, $2b$ and $2L$ respectively. The weight density of the material, which is assumed to behave elastically, is denoted by the Greek letter γ.

The equation relating bending and resistance moments, derived in Chapter 3, can be manipulated to show how the bending moment at mid-span and the section modulus z vary in these two conditions. Units have been omitted for clarity, a permitted liberty when making an algebraic comparison of like with like.

Original size:
$$\text{UDL}(w) = \gamma bd \quad \therefore BM \text{ at mid-span} = \frac{wL^2}{8} = \frac{\gamma bd.L^2}{8} = \frac{\gamma bdL^2}{8}$$

$$\text{and } z = \frac{bd^2}{6}$$

Doubled size:
$$\text{UDL}(w) = \gamma.2b.2d \therefore BM \text{ at mid-span} = \frac{wL^2}{8} = \frac{\gamma.2b.2d.(2L)^2}{8}$$

$$= 2\gamma bdL^2$$

$$\text{and } z = \frac{2b.(2d)^2}{6} = \frac{8bd^2}{6}$$

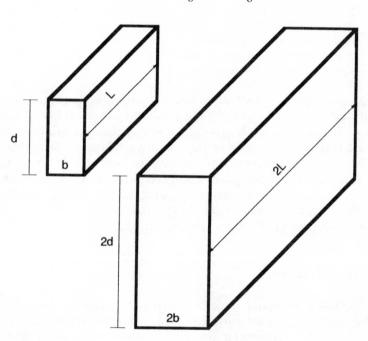

Figure 4.1 *Doubled beam dimensions*

It can be seen that the section modulus, and therefore the strength of the beam, increases by a factor of 8, or 2^3, whilst the maximum bending moment increases by a factor of 16, or 2^4.

Since $BM = fz$ and therefore $f = BM/z$, the stress at the extreme fibres will increase by a factor of $2^4/2^3 = 2$.

It can similarly be shown that if the beam were to treble its size, f would increase by a factor of three, and so on.

There exists a very strong case, therefore, for discarding some of the excess weight as the beam grows in size – a strategy having precedents in nature's devising of the most efficient distribution of matter in vertebrate life forms.

Steel beams and trusses

Rolled steel 'I' section beams

Before investigating the behaviour of trusses *per se,* it is worthwhile examining the reason for the adoption of the steel 'I' section. The necessity for this has been mentioned in Chapter 2, as a consequence of the weight density of steel of 72 kN/m². Given the need to shed some of the unwanted material, the stress diagram in Figures 3.3 and 3.4 suggests that the most important parts of the section are those furthest from the neutral axis. These are preserved in an 'I' section as the tension and compression flanges.

The majority of 'I' section steel beams fabricated in the United Kingdom are known as 'universal beams', although there does exist a range of sections up to a depth of 254 mm still categorized as joists. The deepest and heaviest of the rolled universal beams is the 914 mm × 419 mm × 388 kg/metre section, shown together with its dimensions in Figure 4.2. The tabulated z value of this section is 15616×10^3 mm³. Although the z values of the entire range of manufactured sections are published in a tabulated form, it is essential to understand the logic behind their calculation. Beside the 'I' section is shown its enclosing rectangle, that is the dimensions of a solid section 914 mm deep by 419 mm wide, the z of which is

$$\frac{bd^2}{6} = \frac{419 \times 914^2}{6} = 58338 \times 10^3 \text{ mm}^3$$

The weight of this section in steel would be

(0.914 metres × 0.419 metres × 72 kN/m³) = 27.6 kN/metre run.

The expression $bd^2/6$ for the section modulus emerged from the derivation of the resistance moment of a solid rectangular section. The section modulus of the 'I' section will clearly be much less than this, as can be seen from the tabulated value. An approximate check on this tabulated z value can be made quite simply if it is remembered from Chapter 3 that z is also defined as the first moment of area. This definition is useful when comparatively shallow portions of deep sections are involved, such as the flanges of 'I' beams. The z of the thin web in the zone between the flanges, being rectangular, can be calculated from the formula $bd^2/6$. From the dimensions in Figure 4.2, where A is the area of one flange based on the tabulated dimensions and r is the distance from the centre of the flange to the neutral axis,

$$z \text{ of flanges} = 2A.r = (2 \times 15390 \times 442) = 13{,}604 \times 10^3 \text{ mm}^3$$
$$z \text{ of web} \;\;\;= bd^2/6 = (21.5 \times 799^2)/6 \;\;\;= \underline{2{,}287 \times 10^3 \text{ mm}^3}$$
$$\therefore \text{ total approximate section modulus} \;\;\;= \underline{15{,}891 \times 10^3 \text{ mm}^3}$$

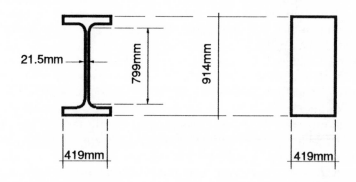

Figure 4.2 *Steel 'I' and solid sections*

This is similar to the tabulated value of $15616 \times 10^3 \text{mm}^3$. The difference arises from the mathematical evaluation of the z of the flanges using calculus, in which the precise Ar value of each successive small thickness of flange is considered.

It is also possible to derive an approximate z value for this section by subtracting the $bd^2/6$ quantities of the two voids in the enclosing rectangle from the z of the rectangular section itself.

A comparison between the 'I' section and a hypothetical solid steel section 914 mm by 419 mm, is a sufficient justification for the adoption of the 'I' beam, i.e.,

	$914 \times 419 \times 388$ kg/m	914×419 solid	% increase
Weight (kN/m)	$(388 \times 9.81) = 3.8$ kN/m	27.6 kN/m	726
Tabulated z (mm³)	$15{,}616 \times 10^3$ mm³	$58{,}338 \times 10^3$ mm³	374

The increase in strength is only about half of the increase in weight.

Castellated steel beams

The steel 'I' section joist and universal beam range are rolled sections without any vertical discontinuity of material. The solid web maintains the lever arm between the centres of the tensile and compressive flanges, and is kept as thin as possible, its thicknesses being conditioned by the need to avoid failure due to shear forces and local buckling. The castellated beam, illustrated in Figure 4.3, is the next stage, although not in a chronological sense, in the transition from a solid section to one with optimum efficiency in bending. By cutting the length of the web of an 'I' section as indicated, and moving the top half of the beam sideways so that the horizontal cuts are aligned and welded, a section with a depth of about one and a half times that of the original is formed. Hexagonal voids appear in the web.

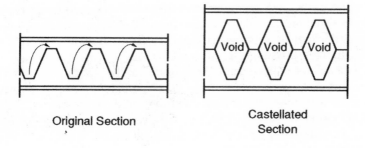

Original Section

Castellated Section

Figure 4.3 *Castellated steel 'I' section*

Taking as the original beam the 914 mm × 419 mm × 388 kg/metre section discussed earlier, the castellation process will increase this to a depth of 1371 mm, with the same breadth and mass per metre length. The tabulated z of this section is 24,168 × 10³ mm³. The approximate z contributed by the flanges alone will be

$$Ar = (2 \times 15390 \times 670.6) \text{ mm}^3 = 20640 \times 10^3 \text{ mm}^3,$$

the difference this time being made up by the second moment of area of the portions of the web above and below the hexagonal voids. The z at this section must be used in design, as tabulated, rather than the greater value for the section between the voids where the web is solid.

Castellated beams are frequently chosen by architects for the visual appeal of the repetition of the hexagonal voids where the beams are not concealed by a false ceiling. This is more likely to apply in the case of roof beams, where a period of fire resistance is not usually stipulated, rather than for floor beams. Where there is a need to allow service pipes to pass at right angles to the direction of the span, the voids, or castellations provide an obvious route, avoiding the need to suspend the pipes underneath the bottom flange, thus taking up more space. In such cases, a false ceiling would normally be provided, unless the architect considered the expression of both structure and services to be of some aesthetic value.

The castellations nearest to the ends of the beams sometimes have to be restored to the solid web state by the welding in of hexagonal plates of the same thickness of the original web. This is occasioned by the high shear forces at the ends of the beam, which may overstress the diminished web of the castellated section.

The steel lattice girder

Whilst the castellated beam pushes the tensile and compressive flanges outwards to create a greater lever arm, leaving a discontinuous web, the lattice girder can be seen as the logical culmination of this process. The parallel flanged lattice girder shown in Figure 4.4 is, unlike the 'I' beam and its castellated alter-ego, formed from totally separate web and flange components. The web, which consists of straight pieces, is connected to the flanges with bolts, welds or a combination of the two.

Several different sectional shapes are used in the fabrication of lattice girders, varying from angles, tees, channels and hollow sections. It has already been demonstrated that tension and compression will exist in the opposite flanges of a beam subjected to bending. It can also be shown that the nature of the force in the diagonals forming the web will alternate between tension and compression. It is therefore possible to manufacture lattice girders as standard components using the least amount of material, so that the compressive members are of a shape conducive to resisting buckling, and the tensile members are just sufficient in cross-sectional area to resist the applied axial forces. There are proprietary systems, on the other hand, which use the same shape of section throughout in the interests of economy in the manufacturing process.

To continue the analogy with the 'I' section, it is possible to estimate the necessary depth of a lattice girder by finding the z required by dividing the permissible stress into the mid-span bending moment, and using the formula $z = 2Ar$.

Thus, if a z value of 15616×10^3 mm^3, i.e. identical to that of the $914 \times 419 \times 388$ kg/m universal beam discussed in the previous two sections, is needed, a flange area can be assumed, and a value found for *r*. It is assumed that the design is to allow for the passage of ducting which could not even be accommodated by a castellated section.

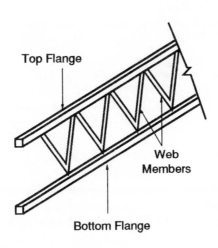

Figure 4.4 *Steel lattice girder*

If each flange is to consist of a square hollow section 150 mm × 150 mm × 16 mm, the tabulated sectional area of which is 66.4×10^2 mm², the expression for *z* can be written as

$$2 \times A \times r = 15616 \times 10^3$$
$$\therefore 2 \times 66.4 \times 10^2 \times r = 15616 \times 10^3$$
$$\therefore r = \frac{15616 \times 10^3}{2 \times 66.4 \times 10^2} = 1176 \text{ mm}$$

Since *r* is the vertical distance from the centre of area of the flange to the neutral axis of the lattice girder, the total depth of the section will be

$$2(1176 + 75) = 2 \times 1251 = 2502 \text{ mm},$$

meaning that the girder will occupy two and a half metres of the height available for the building. If this were found to be unacceptable, a larger flange could be tried, for example a 200 mm × 200 mm × 16 mm square hollow section, which has a cross-sectional area of 117×10^2 mm^2, so that

$$2 \times 117 \times 10^2 \times r = 15616 \times 10^3$$
$$\therefore r = \frac{15616 \times 10^3}{2 \times 117 \times 10^2} = 667 \text{ mm}$$

The depth of the lattice girder would then be

$$2(667 + 100) = 2 \times 767 = 1534 \text{ mm}.$$

This brings the total depth down to about one and a half metres. As the flange area is increased, the value of r decreases, thus reducing the total depth of the structural element.

Given the range of possible shapes for the webs and flanges, and the variations within the equation $z = 2Ar$, it is not surprising that the choice of a particular type and make of lattice girder is sometimes difficult for the architect. Usually, the depth required for lattice girders forming a grid supporting a secondary roof or floor structure is about one-twelfth to one-fifteenth of the span. If the girders are closely spaced to act as a joisted floor or roof system, their depth will be much less, since they will be carrying a narrower band of load than a principal member.

Although lattice girders can be fabricated on site, this is unusual unless the length of span dictates that it must be delivered in two or more sections. For lattice girders, or trusses as they are also known, it has become common practice to use hollow sections for all members because of the comparative simplicity of forming full strength welded end connections. At the other extreme, small span members with light loads often have diagonal web members of continuous steel bars, fillet welded to the flanges. Whatever their composition and whether the spans be great or small, they are still beams or joists which have managed to shed some of their excess weight.

Simple pin-jointed trusses

The basis of the understanding of the way in which forces are transmitted through trusses lies in the principle of the triangle of forces explained in Chapter 1. The simple triangular truss in Figure 4.5 is the fundamental unit of all pin-jointed truss forms, however complex they may at first appear. The term pin-jointed implies that the connection between each pair of sides in the triangular structure will allow a small change in the angular measure as the truss takes up its load. This change will be very small provided that the three members of the truss remain straight. In the truss in Figure 4.5(a), the angle at the ridge will increase, whilst the two angles between the sloping members and the horizontal will decrease.

The sum of these three angles will still add up to 180°. For the purposes of analysis using the principles of statics, these alterations in the geometry can be safely ignored.

To keep the arithmetic and trigonometry as simple as possible, it is assumed that the truss form is that of an isoceles triangle, the angle at the ridge being a right angle. The truss supports a point load of 10kN in this position.

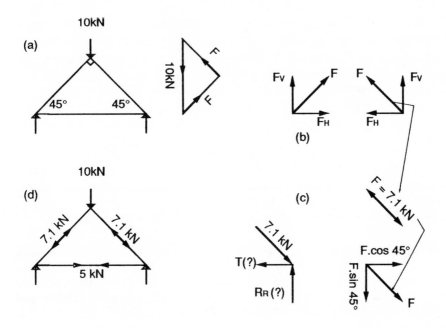

Figure 4.5 *Forces in triangular truss*

The triangle of forces for the three vectors at the ridge can be drawn by using the information given, that is the magnitude and direction of the load and the geometry of the truss. This point, like all points in equilibrium under the action of all of the forces directed towards it, is usually referred to as a node point. The implied logic used here is that since the desired condition of the configuration is that of a stable structure, with equilibrium at all node points, a triangle of forces can be drawn for each of these node points.

Knowing the magnitude and direction of the applied load, and the direction, but not the magnitude, of the forces in the sloping members, the triangle of forces for the ridge node can be plotted. If the 10 kN vector is represented by a vertical straight line to a suitable scale, lines representing the unknown forces in the sloping members can be drawn from its extremities, which will form two of the vertices of the force triangle. The intersection of these two sloping lines will be the third vertex. If the length of these two sides is measured, and converted back into a force by using the chosen scale, the magnitude and direction of the forces in the sloping members is revealed. The direction of the vector arrows, which must follow each other round in the same sense, in this case anti-clockwise, show that each of the sloping members has an upward vertical component. This is to be expected, since these two members are supporting the 10 kN load at this point.

These forces can, alternatively, be directly calculated by expressing the forces in the sloping members in terms of their components, as described in Chapter 1. In this symmetrical truss, the two vertical components will have the same magnitude, and together they will equal the applied 10 kN force. If the unknown forces are denoted by F kN then

$$2 \times F \times \sin 45° = 10$$
$$\therefore 2 \times F \times 0.707 = 10$$
$$F = \frac{10}{2 \times 0.707} = 7.1 \text{ kN}$$

This calculation reveals that the resultant force F in each of the sloping members is 7.1 kN. To verify this result, F can be resolved into its vertical and horizontal components F_V and F_H as in Figure 4.5(b), i.e.

$$F_V = F.\sin 45° = (7.1 \times 0.707) = 5 \text{ kN} \quad \text{and}$$
$$F_H = F.\cos 45° = (7.1 \times 0.707) = 5 \text{ kN}$$

The vertical equilibrium at this node point, $\Sigma V = 0$, can be expressed by the convention

$$(5 \text{ kN} \uparrow) + (5 \text{ kN} \uparrow) = 10 \text{ kN} \downarrow$$

meaning that the downward applied force of 10 kN is balanced by the two upward forces of 5 kN.

Similarly, the horizontal components of the two forces F are in equilibrium, i.e.

$$5 \text{ kN} \rightarrow = 5 \text{ kN} \leftarrow$$

thus fulfilling the equilibrium condition $\Sigma H = 0$.

The third of the equilibrium conditions, $\Sigma M = 0$, is of no relevance here, since there will be no moments at the ends of members which are pin-jointed.

The force triangle also divulges the nature of the forces in the sloping members. The vector arrow associated with the ridge node point is directed outwards from the member in each case. In keeping with the logic of the sign convention adopted in Figure 2.2, the force in each member must be described by a second vector arrow pointing in the opposite direction. The vector notation for each of these members is shown in Figure 4.5(d), revealing them both as struts.

The vector arrow at the lower end of each of these members can now be used to evaluate the magnitudes of the reactions and of the force and its nature in the horizontal member. Figure 4.5(c) shows the known forces directed towards the node point at the right-hand reaction. By the same procedure adopted for the ridge, the 7.1 kN vector can be drawn to scale, and the known alignment of the vertical reaction and horizontal vectors used to complete the force triangle. The magnitudes of the two unknown forces can be found by scaling as before. The upward direction of the reaction vector is a predictable outcome, from the point of view of both intuition and of Newton's Third Law. The inward pulling vector on the horizontal member demands a complementary inward vector at the other end, showing this member to be in tension. These vector arrows are shown in the diagram of the solved structure in Figure 4.5(d).

As before, these forces can also be calculated from the vertical and horizontal equilibrium conditions. Calling the right-hand reaction R_R and the force in the horizontal member T,

$$\text{For } \Sigma V = 0, \quad R_R = 7.1 \times \sin 45°$$
$$= 7.1 \times 0.707$$
$$= 5 \text{ kN.}$$

$$\text{For } \Sigma H = 0, \quad T = 7.1 \times \cos 45°$$
$$= 7.1 \times 0.707$$
$$= 5 \text{ kN.}$$

The figures appear simpler than those for the ridge because they just involve resolving F, the applied force from the point of view of the node point when using this sequence of calculations, into its two orthogonal components. The directions of the vector arrows can also be discovered in this way, i.e.

Vertical reaction = Vertical component of F (↑5 kN = ↓ 5 kN)
Force in horizontal member = Horizontal component of F (← 5 kN = → 5 kN)

The pitched roof

Evolution of pitched forms

'There is, in cold countries exposed to rain and snow, only one advisable form for the roof-mask, and that is the gable, for this alone will throw off both rain and snow from all parts of its surface as speedily as possible.' John Ruskin, in *The Stones of Venice,* although referring to the pitched roof as a seminal influence in the evolution of the Gothic style, was also drawing an important comparison between structural form and climatic conditions. The further north one travels, the steeper becomes the pitch of the majority of the domestic roof structures. The flat roof, whilst being admirably suited to a Mediterranean climate, can, if not constructed to a standard approaching perfection, collect a permanent pond of water if an unintended slope should develop in the wrong direction. This is not the place to try to resolve the eternal debate as to the relative merits of the flat and the pitched roof. The issue is a complex one, and the opinions held by environmental scientists, structural engineers and art historians, not to mention architects, differ widely. In terms of pure structure, the inescapable truth is that the pitched roof generates its own structural depth, the depth at mid-span of the truss form increasing with the angle at which the roof is pitched. This is not to say that the roof will be any cheaper in terms of material cost or labour. The reverse is often the case. Cost alone, however, would not necessarily be a prime consideration when deciding between two such contrasting architectural statements.

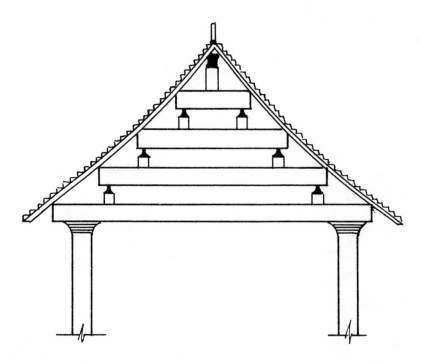

Figure 4.6 *Traditional Chinese pitched roof*

The basic triangular form analysed in the previous section is an implicit intention whenever an architect proposes a pitched roof. This has not always been the case in all cultures, however. The pitched roof form in traditional Chinese construction was achieved with a trabeated system, that is a system of horizontal beams. As shown in Figure 4.6, the length of successive beams was conditioned by the pitch of the roof, each shorter one transferring its reactions to the one below as a pair of point loads, causing bending moments in the longer beam. The brackets would usually be decoratively carved, since this was a roof form where structure was openly expressed.

The primary design intention in all triangulated pitched roof forms is to receive the loading from the secondary members, such as common rafters and purlins, in the form of point loads applied at the node points. This arrangement in a truss will result in only an axial tensile or compressive force in each member. The resistance moment of the truss system is generated by the interaction of these forces and the lever arms associated with their lines of action. There will be no bending moments in the individual members, apart from those arising out of their self weight, which will be comparatively small. This condition cannot always be brought about, however, in particular in the context of modern domestic roof construction in the United Kingdom, for two reasons.

First, when, as is usually the case, the horizontal tie member of a truss also forms the structure of the ceiling, this member will try to support the weight of the ceiling and the inevitable contents of the roof space by spanning between the nearest available points of support, whether these be node points in the truss or internal walls within its span. Code of practice BS 5268:PART 3:1985 stipulates a minimum uniformly distributed load for which the combined ceiling joist and tie must be designed. The combined stress condition has to be carefully checked. It can be shown that where a member subjected primarily to an axial tensile stress is subjected additionally to a bending moment, there will be an increase in that tensile stress. The axial force will have become eccentric, the consequence of which will normally be an increase in the required size of the tie.

The second occurrence, equally common, of a combined bending and axial stress condition, arises from the adoption of the trussed rafter system for domestic pitched roof construction. Here, the trusses, the most usual spacing of which is 600 mm, are self-sufficient as far as the vertical loading is concerned, acting in much the same way as a series of parallel floor joists. Unlike, therefore, the traditional system of roof construction in which the common rafters are supported by purlins, which in turn are supported at the node points of the main truss, every inclined rafter forms part of a triangulated truss – hence the terminology of 'trussed rafter'. Each rafter in this system, moreover, receives its load directly from the tiling battens or the plywood roof sheathing as a uniformly distributed load. Bending moments are therefore caused in the rafters in addition to the axial compressive force associated with the triangulation of the truss.

Internal triangulation in pitched roofs

Although the simple triangular truss is well suited to the support of a single point load from a ridge purlin, only a pitched roof spanning a very small distance would be structured in this way. A traditional roof would normally incorporate purlins which have to be supported somewhere along the sloping rafter of the main truss. The purpose of these purlins is to reduce the span of the common rafters, which would be needlessly deep if required to span from eaves to ridge.

Figure 4.7(a) represents a pitched roof truss in which the loads from the common rafters are transferred to it at three purlin positions, one at the ridge and one at mid-length of either rafter. No magnitudes have been affixed to the vector arrows representing the point loads from the purlins, the purpose of this exercise being solely to demonstrate the direction of the forces in the internal members. It is assumed that all three purlin loads are equal.

The spaces between the members and reactions have been described by capital letters in accordance with Bow's Notation, a graphical construction devised by Robert Bow in 1873. A vertical line representing the load and reaction vectors, using the corresponding lower case letters, can then be drawn to scale. By constructing force triangles and polygons for each node point in turn, the force diagram for the complete structure in Figure 4.7(b) evolves. From this, the individual force diagram for any node point can be extracted, thus revealing the magnitude and direction of the force in each of the members meeting there. That for *BCGF* contains the vector *bc* for one of the purlin loads, the direction of which is, by definition, downwards. This means that this vector arrow is running clockwise relative to the closed polygon, so that the other three arrows must also run clockwise. This reveals that *cg* and *fb* are pushing towards the node point, confirming the intuitive feeling that members *BF* and *CG* are in compression. The arrow for vector *gf* is also directed towards the node point, indicating that member *GF* is also a strut.

Similarly, the force diagram for the node point *CDIHG*, geometrically not a polygon but a triangle attached at one vertex to a quadrilateral, but still a valid reflection of equilibrium at a point, can be drawn. The load vector *cd* is known to point downwards, and the arrows for the remaining four vectors can be inserted head to tail. For this type of force diagram, not being a single plane bounded by straight lines, the direction changes from clockwise in the triangle to anti-clockwise in the quadrilateral, but this is only a consequence of two of the vectors being common to both

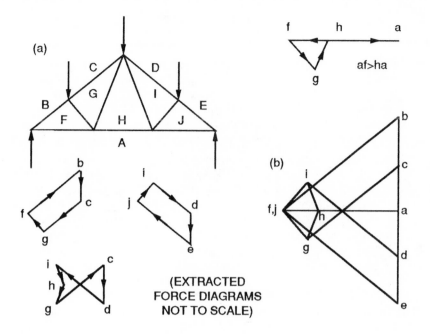

Figure 4.7 *Bow's Notation*

constituent polygons. The essential feature of their reflection of an equilibrium condition is that the head of one arrow follows behind the tail of the next. The outcome of most interest here is the revelation of members *IH* and *HG* as ties, since vectors *ih* and *hg* respectively pull away from the node.

The force diagram at the node point *FGHA*, which is in the shape of a triangle with one side produced, or extended, confirms that *GF* is a strut and that *HG* is a tie. Both have horizontal components acting from left to right, accounting for the tensile force in member *AF* being greater than that in member *AH*. What is important to notice here is the equality of the vertical components in these two members, so that the downward push from *GF* is exactly balanced by the upward pull from *HG*. This is an essential feature of the behaviour of triangulated systems, in that the horizontal tie is relieved of any residual vertical force at this point, the presence of which would have created considerable bending moments, causing an increase in the required depth of the tie.

Since the truss is symmetrical in both its loading and its geometry, the force diagrams for the node points to the right of the centre line will be mirror images of those on the left. Therefore *IH*, like *GH*, is a tie and *JI*, like *GF*, is a strut.

The function of the internal members, therefore, is to provide a load path to the supports for the two intermediate purlin loads without inducing bending moments in the rafters with which they coincide. Whilst a part of each purlin load *BC* and *DE* travels straight down the lower half of the external members, that is the rafters, the greater part of these forces is transmitted by strut action to the bottom tie, back up to the ridge by tie action, and finally down to the supports via the rafters. The exact proportion of load transferred in this way depends on the slopes of the rafter and of the internal strut.

This particular truss configuration lends itself to that sense of structural economy whereby the longer members are in tension, and the compressive forces are reserved for the shorter members which are less prone to buckling. Exposed trusses of this form can be a source of expression of structure in architecture, in that those members corresponding to *IH*, *HG*, *AF*, *AH* and *AJ* in Figure 4.7 can be of minimal cross-sectional area, contrasting with the more bulky compressive members.

Vertical struts and ties

As a corollary of the arguments set forth in the previous section, the nature of the forces in all of the members in Figure 4.8(a), in which there are two purlin loads on either side of the ridge as well as the one in the centre, can be deduced without further analysis. The same pattern of load transference will create two additional sets of forces, each one consisting of a strut and a tie in a 'V' formation with equal vertical components. The horizontal member joining the supports remains in pure tension under the action of the five point loads. The appropriate vector arrows are indicated on the line diagram.

Where the span of the truss is short and the pitch steep, one of the arrangements in Figures 4.8(b) and 4.8(c) may be preferred. In 4.8(b), the struts beneath the two purlins either side of the ridge are in a vertical position, whilst the two corresponding ties remain inclined. In 4.8(c), the struts are inclined as before, but the two ties merge into a single vertical member. The tensile force in this member will balance the sum of the vertical components of both struts where all three members meet the horizontal tie. It is a common error to imagine the vertical member in this configuration to be in compression owing to its position immediately under the central point load. This load, as in the case of the truss analysed in the previous section, travels straight down the inclined rafters.

In Figure 4.8(d) the vertical member extending down from the ridge point acts as a tie, but does not take any part in the triangulation of the forces in the truss. At its lower end it is not possible to draw the closed polygon compatible with an equilibrium state, since the vertical force would remain unresolved. The horizontal tie member cannot be resolved in such a way that it possesses a vertical component, and cannot therefore balance the sole vertical force vector directed away from this node point. The function of an unresolved vertical tie such as this is merely to alleviate the sagging which might otherwise occur in a long lightly stressed horizontal tie member.

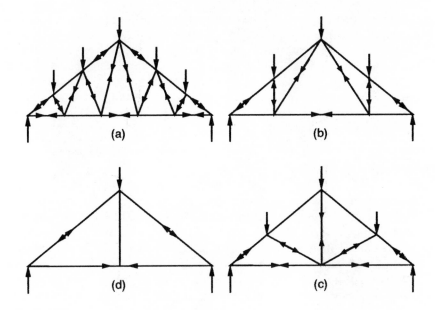

Figure 4.8 *Forces in pitched roofs*

A truss form where compression occurs in a central vertical member is illustrated in Plate 8. Here, intermediate supports are provided for the rafters, some of whose load is transferred to the crown post, as this central strut is known. The vertical force vector pushing downwards onto the horizontal tie is still unresolved in terms of triangulation. It is resolved, however, by the shear force on either side of the point of application of this force. The horizontal member is therefore acting as a beam supporting a central point load in addition to its function as a tension member. The shear force and bending moment diagram for a beam supporting a central point load appear in Chapter 1, to which the behaviour of this horizontal member in bending can be compared. The description of such members as 'tie beams' in medieval architecture is appropriate, with hindsight, to what is in effect an eccentrically loaded tie. The eccentricity here is caused by a varying bending moment.

Plate 8 *Crown Post Truss*

Trusses with raised ties

All the pitched roof trusses discussed so far in this chapter have tie members coinciding with the horizontal line joining the supports. In all these forms the structural depth of the truss, taken as the altitude of the triangle at mid-span, where the bending moment on the system is greatest, is determined by the pitch of the rafters. From Figure 4.9, it can be seen that this altitude is equal to half the span of the truss multiplied by the tangent of the rafter pitch. Dividing this into the span L, the span to depth ratios, based on the line diagram of the truss configuration, can be tabulated as follows:

Pitch of truss	Tan (α)	Altitude	Span to depth ratio
15°	0.268	0.134L	7.5
22½°	0.414	0.207L	4.8
30°	0.577	0.289L	3.46
45°	1	0.500L	2
60°	1.73	0.867L	1.15

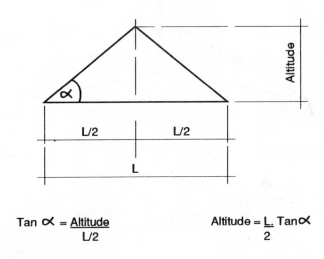

$$\text{Tan } \alpha = \frac{\text{Altitude}}{L/2} \qquad\qquad \text{Altitude} = \frac{L.\text{Tan}\alpha}{2}$$

Figure 4.9 *Pitched roof dimensions*

Since it is usually possible to devise a truss form in which the member sizes are small relative to the space enclosed by its peripheral members for span to depth ratios of up to 15, even the shallowest pitched roof trusses are not found lacking in their structural depth. For the steeper pitches, the main difficulty is likely to be the length of the rafters and of the internal compression members.

It is sometimes proposed, for the purpose of either structural expression or of increasing the ceiling height near the centre of a space, to construct the bottom member as two inclined ties intersecting at mid-span. Figure 4.10 shows two such forms of 'scissor truss' as they are sometimes known. In 4.10(a), where the configuration is for a single point load at the ridge, a vertical member is introduced under this point. The force triangle at its lower end reveals that this member is a tie – a condition intuitively deduced from its role in preventing the two bottom tie members from reverting to a straight line. At its upper end, the downward pull which it exerts on the ridge node point increases the magnitude of the vertical components in the rafters, and as a consequence the resultant force therein. This is to be expected, because the reduced structural depth of the truss must be compensated by higher forces in the resistance moment couple of the system.

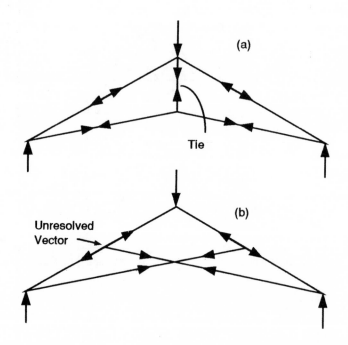

Figure 4.10 *Raised ties in pitched trusses*

The form in Figure 4.10(b) is not a fully triangulated system, since the uppermost of the three spaces is a quadrilateral. The two inclined tie members are extended to meet the rafters, causing an unresolved inward pull on both of them. The consequences, therefore, of reducing the structural depth in this case have been to induce bending moments as well as compression in the rafters. If this configuration were pin-jointed at every node point, the lack of triangulation in the upper part of the truss would cause excessive changes in geometry, or even failure. This is prevented from happening by the continuity of the rafters for their entire lengths between eaves and ridge.

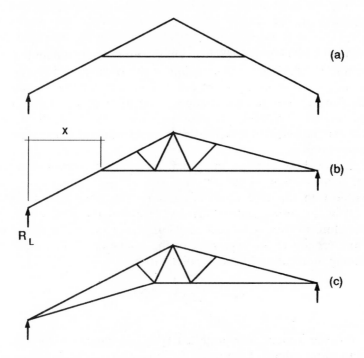

Figure 4.11 *Incomplete triangulation in trusses*

Collared trusses

Figure 4.11(a) shows a truss totally unlike any of those previously discussed so far in that the outward horizontal component of the compressive force in each of the rafters is not resolved at the supports. There is no tie member at these positions to prevent the rafters from spreading. It looks at first sight as though the horizontal tie has simply been moved about two-thirds of the way up the rafter, probably to increase the headroom in the space below, with a consequent reduction in the structural depth. But when it is observed that the rafters, whether the roof loading is in the form of a uniformly distributed load or of a point load, will deflect inwards, it is clear that the raised horizontal member must shorten, albeit to a very small extent. This member, or collar as its position defines it, is therefore a strut and not a tie. Even if the collar were designed to safely resist this compressive force, this system should be avoided except in the case of very small spans, since its stability depends on the capacity of the supporting walls to resist the unresolved outward force component, more commonly called the thrust.

The extended rafter

Figure 4.11(b) depicts a form frequently proposed in modern domestic architecture in which the triangulation of the pitched roof is interrupted on one side for a relatively short distance to accommodate a window in the rafter plane. Any sudden change in the form of a structure is to be avoided wherever possible, but an overriding need for light in this position can be satisfied provided that two structural consequences are fully understood.

The first of these rests on the premise that the thickness of a wall in domestic construction, about 200 mm with an intervening cavity, is not compatible with the acceptance of primary horizontal forces at its top. The wall at the untriangulated left-hand side of the roof structure can only therefore provide a vertical reaction, R_L. If the horizontal distance from this reaction to the beginning of the triangulation is x metres, then the bending moment at this node point will be

$$R_L x - \frac{wx^2}{2} \quad \text{kN-m}$$

where w is the horizontally projected UDL on the extended rafter. This will result in the whole of the rafter section being deeper than had it been solely in compression. It is not sufficient merely to provide a deeper section just for the extended part of the rafter. This junction is by necessity a moment connection (see Chapter 5), and it is only theoretically possible to discontinue this enlarged section at the next node point up the rafter.

The second consequence of adopting this form concerns the connection between the walls and the rafters. Although the tendency for the walls to be pushed outwards is not so marked as in the case of the collared truss with its greater length of unsupported rafter, there will still be a small horizontal movement as the roof structure enters its equilibrium state. A detail whereby the truss is seated on top of the wall, whilst still restrained against wind uplift, lends itself more easily to the accommodation of this tendency to slide, as compared with a joist hanger connection onto the inside face.

A better structural solution is shown in Figure 4.11(c), in which the left-hand portion of the horizontal tie is inclined downwards to the support, completing the triangulation and relieving the walls of any thrust.

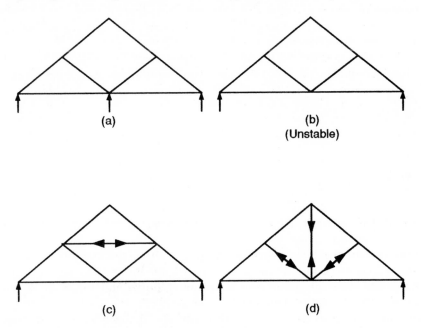

Figure 4.12   *Internal arrangements in pitched roofs*

Internal wall supports to pitched roofs

In a roof of traditional construction, the load path from the tiling battens, common rafters, purlins through to the main trusses is easy to follow. This is especially so if a distinction is made between the common rafters and the principal rafter of each truss.

This is not always the case. It was common practice in English house building in the late nineteenth and early twentieth centuries to support every fifth or sixth common rafter with an inclined strut, and to transmit some of the load from the remaining rafters to it by means of a purlin set beneath the rafter plane. The pair of inclined struts would meet at a ceiling joist, where they would be supported on an internal wall. This spurious form of truss, as shown in Figure 4.12(a), actually consists of two trusses, each one a simple triangle, one end of each resting on the internal wall. These support the upper part of the rafter which, in the absence of a tie member at the level of the purlins, imposes both vertical and horizontal forces on to the apex of each one. The net result of this was that the internal wall supported about as much weight as the two external walls combined.

It would not be possible to use this truss in a pin-jointed form without the help of the internal walls, as in 4.12(b), because of the unstable central quadrilateral. This becomes possible only with full triangulation. This could be, and possibly has been on occasion, restored as in 4.12(c) by the addition of a horizontal strut linking the two purlins, creating a structural form consisting of a triangle resting on a parallel flanged lattice girder. Great care should be taken in rehabilitation work when contemplating the removal of such an apparently redundant member – its necessity may be discovered with some alarm at too late a stage.

Figure 4.12(d) shows the triangulation restored by the insertion of a vertical member, which becomes a tie in a truss identical to that in Figure 4.8(c). Its upward vector usurps the function of the internal wall.

Connections in trusses

Node points

'In all the mechanical side of anatomy nothing can be more beautiful than the construction of a vulture's metacarpal bone.' Sir D'Arcy Wentworth Thompson's observation in his seminal work *On Growth and Form* draws an important analogy between the construction processes of nature and those of the architect and engineer. The triangulated form of this bone has a configuration identical to that of the trusses analysed earlier in this chapter, the internal spaces being isosceles triangles. It is reasonable to assume, since vultures do not have noticeable difficulties in flying, that nature was seeking to devise a structural form capable of spanning a considerable distance using the least amount of material.

Leaving aside the obvious variations between buildings and vultures, the structural feature which most markedly differentiates the engineer's truss from the metacarpus is the way in which the linear components are joined together. The structure of the vulture's wing is organic, in that the linear members of its truss continue smoothly from one to another. Nature saw no reason to pause and establish a discontinuity, thus creating the need for a nodal connection. Nature has enough problems transferring forces from bone to muscle with delicately connected tendons. She does not need to create problems for their own sake.

The engineer, on the other hand, finds the truss, or open-webbed beam, more conducive to fabrication from finite lengths of materials such as timber or steel. The nodal connection therefore manifests itself as a separate component whose function is to receive the ends of those struts and ties intersecting at that point. In terms of structural logic, what it is receiving are forces, whose components it reorientates from one member to another according to the laws of static equilibrium.

Ideally, the nodal connections in a truss should be such that all of the forces which keep it in equilibrium act in the same plane. This is not difficult to achieve in a steel truss, where a gusset plate made of the same material is welded or bolted to each of the incoming angle or channel sections. No material is involved other than that of which the constituent members are made, so that the complete truss is in a sense organic. The nature of the connections, however, is such that designers can safely assume that any bending moments that arise at the ends of the struts and ties as the truss takes up its load and tries to adjust its geometry are negligible. Although a theoretically perfect pin joint can only be achieved with a frictionless single bolt connection through each member into the gusset plate, these secondary bending moments arising from the fillet welds or multiple bolt arrangement are usually of a very small order.

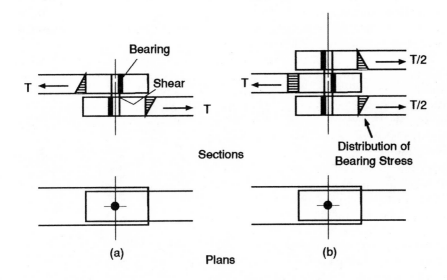

Figure 4.13 *Bolted connections in timber*

Timber truss connections

A modern truss made of timber struts and ties does, on the other hand, require the intervention of a different material for a node point to evolve. The dovetailed and tenon joints used in traditional carpentry have now disappeared from structural design in favour of the steel bolt, the exception being in the deliberate reconstruction of historic buildings. The overriding problem in transferring a force from one timber member to another through a steel connection is the relative strengths of the two materials. Figure 4.13(a) shows the path taken by a tensile force T through a simple lap connection involving a steel bolt. This bolt will eventually transfer this force by its shear strength, which is proportional to its cross-sectional area, in the same way as a bolt on a steel truss. But this force must first enter the bolt by the direct bearing between the timber and the comparatively small bolt. This can only take place over an area bounded by the diameter of the bolt and the thickness of the timber.

The bearing stress caused at this interface, whilst well within the capacity of the steel, would, for all but the smallest trusses, exceed the much smaller bearing capacity of the timber. It is therefore necessary to increase the width of timber over which this bearing load is transmitted. This is achieved by the inclusion of a steel connector such as a split ring or a toothed plate, greater in diameter than the bolt, but held in position by it. This increases the contact area, thus reducing the bearing stress. These connectors, being located at the interfaces of the connected members, are hidden from view, only the bolt being visible.

Because this connection method displaces the member force laterally, that is into another plane, a feature of large span timber trusses is the complex arrangement at the nodes. The single lap joint is usually avoided, since the eccentricity of the two forces can cause excessive stresses on the inner face of the timber. It is always preferable to join one member to two, or two members to three, thus conserving the alignment of the forces. This type of connection, shown in Figure 4.13(b), divides the force T in half, so that two sections of the bolt are in shear. The bearing stresses on the central member, which carries the whole force T, are evenly distributed over the length of bolt with which it is in contact. The uneven bearing stresses occur only in the two outer members, to each of which is transmitted a force of $T/2$. As well as providing data on the safe loads which can be safely transmitted across these components, code of practice BS 5268:PART 2:1984 stipulates the minimum edge and end distances to the centre of a bolt from the respective faces of the timber. The edge distance is critical because the force in the member must first travel around the sides of the connector to its remote face before the bearing stress comes into play. If the end distance is insufficient, the connector could be pushed between the longitudinal fibres of the timber in the direction of the grain.

Timber trusses can often be made aesthetically satisfying by emphasizing rather than trying to conceal the contrast with the visible steel nodal connections – a trait even more noticeable in three-dimensional forms where the connected members are bolted through to a steel plate.

At the other extreme lies the trussed rafter which, by virtue of its close spacing, is lightly loaded in comparison with the more widely spaced principal truss. Trussed rafters consist of single struts and ties all of which lie in the same plane. Here, the member forces are transmitted to thin steel plates, not by the introduction of separate components such as nails or screws, but by toothed projections punched out of the plates themselves. These are hydraulically pressed over both sides of the members meeting at the node, so that the permissible loads are still conditioned by the bearing strength of the timber and the resistance to deformation of the teeth.

Reinforced concrete trusses

The closest simile that man has devised for the organic wing structure of the vulture is the triangulated reinforced concrete truss. This is a rarely used system, however, for that very reason. The continuity of form at the node points will attract secondary bending moments which cannot be ignored, especially if the strut and tie dimensions are conditioned by a large span and heavy loading. It is, ironically, these very conditions which are likely to lead to the selection of this form, with its advantage of fire resistance not possessed by unprotected steel. A steel truss, if required to be clad in concrete, may well be more expensive and just as time-consuming as a similar form in reinforced concrete. Formwork and concrete mixing plant are, after all, necessary in both cases.

The reinforced concrete truss comes into its own as a Vierendeel girder, a structural form named after the Belgian engineer, Professor Arthur Vierendeel, who devised it in 1896. Here, the diagonals are omitted and replaced by vertical members which are rigidly connected to the main horizontal struts and ties. The elevation now consists of rectangles instead of triangles. These are liable to distortion if the nodal connections are pinned. The absence of a diagonal member means that opposite corners of the rectangle are free to approach one another if that diagonal were a strut, and to recede if it were a tie. This potential distortion is prevented by the rigid connection between each vertical and horizontal member, thus making sure that the right angle remains a right angle. Bending moments are consequently attracted to the ends of these members, giving rise to a condition for which reinforced concrete may well provide the optimum solution.

It is, again with a sense of irony, inside a solid reinforced concrete beam that truss action provides a design theory for coping with shear forces. The 'web' is postulated as a series of diagonals, alternately in tension and compression. The compressive forces can be resisted by diagonal bands of concrete, whilst the tensile forces are provided by inclined steel bars aligned with their lines of action.

Chapter 5

PORTAL FRAMES AND ARCHES

Portal frames

One dictionary definition of the word 'portal' is 'a gate or doorway, especially a great or magnificent one'. This immediately suggests an architectural feature of great visual significance in the elevation of a building, such as the aedicule, that is two columns supporting an entablature and a pediment. The term 'portal frame', on the other hand, when used to describe a structural element in a building, implies the existence of a quality wholly lacking in an architectural form created from distinct elements carved out of stone. This quality is that of continuity between vertical and horizontal members, in this case between beam and column.

A building structured on the principle of the portal frame may not necessarily express this principle either externally or internally. Unless the architect is at pains to make strong visual statements using either the geometry of the frame's outline, or the continuity of the beam into the column, or indeed both, the adoption of the portal frame will remain simply an engineering response to a structural problem, and will have little or no significance for the architecture.

The idea of continuity

Tree structures

Continuity between two structural elements can best be observed and understood with reference to a simple form consisting of just one beam and one column. Such a form is manifested in a lamp post with a light on one side only, depicted in Figure 5.1(a) with the vertical uniformly distributed load, *w* kN per metre run, representing the self weight of the beam together with any mandatory live load caused by the build-up of ice on the surface. If the length of this horizontally projected beam is *L* metres, the bending moment caused by this load at the junction of the beam and the column will be the algebraic sum of the moments about that point. Since only one load is involved, there will be only one term in this expression, i.e.

$$BM = wL \times \frac{L}{2} = \frac{wL^2}{2} \text{ kN-m}$$

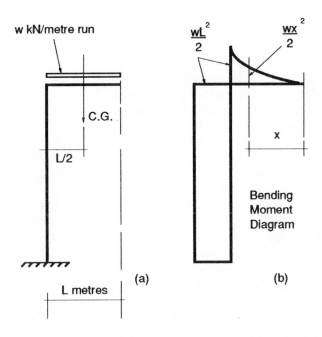

Figure 5.1 *Bending moments in 'tree structure'*

Clearly, the horizontal portion of the structure cannot simply rest on the column. The bending moment at its support cannot just vanish. Unless separation between the two portions of the structure is to occur, the resistance moment possessed by this end of the beam must also be possessed by the top of the column, without any separation between the two. This allows the bending moment to be carried round from one member to the other.

To draw the bending moment diagram for the complete form, it is essential to recall the identity between a bending moment diagram and a graph, in that the ordinate representing the bending moment value at a particular point is plotted at right angles to the line representing the axis of the member. The set of bending moments for the column can therefore be plotted towards the left of this line. The magnitude of these bending moments will be the same for every point on the column, and also equal to that at the junction between the two members, since all of these points are equidistant from the centre of gravity of the load wL.

The value of the bending moment at any distance x from the end of the beam will, again by reference to the definition of bending moment, be the product of the load to the right of that section and the distance from its centre of gravity to that point, i.e.

$$BM_x = wx \times \frac{x}{2} = \frac{wx^2}{2} \quad \text{kN-m}$$

As shown in Figure 5.1(b), the value of the bending moment increases as the square of the distance from the unsupported end, giving the graph an upward parabolic curve as its path is traced to the left. This graph is similar to that part of the parabola in Figure 1.7 which lies to the left of the vertical axis. In this case the equation of the curve is $y = Ax^2$ where $A = w/2$. Even though x is negative in this zone of the parabolic function, y will still have a positive value since the square of a negative quantity is positive, i.e.

$$(-x)^2 = x^2$$

An important feature of the bending moment diagram in Figure 5.1(b) is that it has a non-zero value where the structure terminates in the ground. This implies that the foundation has to be treated as a downward extention to the structure, not only as a recipient of a vertical force, but as a means of resisting an overturning effect. The foundation is therefore eccentrically loaded. Of particular relevance to the topic of portal frames, however, is that the junction between the column and the ground is fixed, so that no rotation between the two is permitted. This condition of fixity, denoted in Figure 5.1(a) by the hatched lines, is analogous to that of the unopenable door discussed in Chapter 1.

Hinged and fixed base connections

The structural form in Figure 5.1 could not be constructed in any other way than with a fixed base. If the moment at the base of the column were not resisted, the structure would collapse by rotation about the base. The connection giving rise to this mode of failure can be described as hinged, in keeping with the analogy of the opening door in Chapter 1, or pinned. The notation for these two extreme theoretical conditions is shown in Figure 5

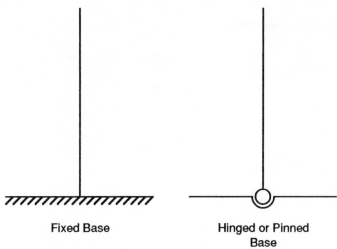

Fixed Base   **Hinged or Pinned Base**

Figure 5.2 *Column base notation*

The contrast between the two is expressed quite dramatically by the process of cutting down a tree, which in its stable condition, by virtue of its botanical evolution, has a fixed base. The removal of a wedge of timber near to the base of the trunk reduces the structural depth and the lever arm, until the resistance moment of the remaining section is so small that the trunk begins to rotate. A fixed base has been converted to a hinged base. If the wedge is cut from one side, the eccentricity of the load from the upper part of the trunk will cause it to rotate towards that side. For trees of wide girth, different sized wedges are cut from both sides so that the sawyers are standing on the side of the tree away from the direction of rotation at the moment of failure. Whether one uses experience, instinct or the laws of structural mechanics, it is as well to be totally certain as to this direction before starting to fell a tree. He who judges incorrectly may emerge flatter from the experience.

All structural forms, whether natural or man-made, in which all the load is channelled down a single vertical support, can therefore conveniently be classed as 'tree structures'. The base must be fixed. Where there are two supports, as in Figure 5.3, resistance to overturning is generated by the horizontal distance between the bases. This is demonstrated in the calculation in the next section. The columns may in this case be connected to the bases either by hinged or by fixed connections but, for ease of analysis, hinged bases are used. The third hinge at the mid-span point of the beam defines the system as a three-hinged portal frame, a form for which the forces and bending moments can be determined using the laws of statics – hence the description of 'statically determinate'.

Three-hinged portal frames

Lateral loads

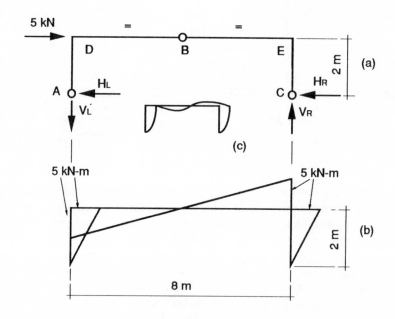

Figure 5.3   *Three-hinged portal frame – lateral load*

The three hinge points in Figure 5.3(a), denoted as *A, B* and *C,* will, by definition, be positions of zero bending moment under all conditions of loading. By equating the clockwise and anti-clockwise moments about any of these points, therefore, the magnitude and direction of the unknown external reactions can be determined. To write these equations, assumptions have to be made as to the probable directions of these reactions. In making these assumptions, the equilibrium conditions $\Sigma V = 0$ and $\Sigma H = 0$ should be kept in mind.

The horizontal point load at the eaves position in this example, although an apparent abstraction, does in fact depict a possible load path for wind forces through a portal framed structure. The point load would occur in this position when secondary vertical members between the portal frames were restrained at their upper ends by an edge beam, which in turn would be loaded horizontally. The reactions to this edge beam would then impose horizontal forces on the portal frames.

The assumed direction of both horizontal reactions is towards the left, which is reasonable since the point load is acting towards the right. The two vertical reactions must act in opposite directions to maintain the $\Sigma V = 0$ condition in the absence of any other vertical forces in the system. The left-hand reaction is assumed to act downwards, and the right-hand reaction upwards. It will be seen in the analysis that this problem is an example of the way in which vertical reactions can be generated by the application of a purely horizontal force.

V and H denote vertical and horizontal reactions respectively, whilst L and R as subscripts distinguish between left and right.

If B is a hinge in a structure in equilibrium, the sum of the moments taken about it from either side must be zero. The 5 kN point load will not appear in these equations because its line of action passes through B.

$$\text{From left-hand side,} \quad \Sigma M = 0, \therefore 4V_L = 2H_L$$
$$\text{From right-hand side,} \quad \Sigma M = 0, \therefore 4V_R = 2H_R$$

Since $\Sigma V = 0$, $V_L = V_R$, so that $2H_L = 2H_R$ and $H_L = H_R$; and since $\Sigma H = 0$, $H_L + H_R = 5$, and using the equality of these two horizontal reactions proved in the previous statement,

$$H_L = H_R = \frac{5}{2} = 2.5 \text{ kN}$$
$$\text{Since } 4V_L = 2H_L, \quad V_L = \frac{2 \times 2.5}{4} = 1.25 \text{kN} \downarrow$$

Since V_L has emerged with a positive value, its assumed downward direction was correct.

Also, since $\Sigma V = 0$, V_R must have the same magnitude, but must act in the opposite direction, that is upwards

$$\therefore V_R = 1.25 \text{ kN} \uparrow$$

The assumption that both horizontal reactions act towards the left is therefore correct. The intuitive feeling that the horizontal reactions would each be equal to one half of the point load has also proved to be correct. This is only the case, however, when the hinge position *B* is at mid-span, as in this example, making the frame geometry symmetrical. As point *B* moves to the right, a greater proportion of the horizontal point load is transferred to the hinge at *A,* the left-hand reaction.

The bending moments at the beam to column junctions, denoted in Figure 5.3(a) as *D* and *E,* can be calculated solely from the horizontal reactions, since the lines of action of the vertical reactions and of the point load pass through those points, and therefore exert no moment about them. Each of these junctions must exhibit the continuity described in the previous section, since it has to transfer a bending moment between the two members meeting there. These junctions are known as rigid joints.

Because of the symmetry of the frame and the equality of the horizontal reactions, the bending moments at *D* and *E* will have the same magnitude, i.e.

$$BM \text{ at } D = (H_L \times 2) = (2.5 \times 2) = 5 \text{ kN-m}$$
$$BM \text{ at } E = (H_R \times 2) = (\ 2.5 \times 2) = 5 \text{ kN-m}$$

The ordinates of the bending moment diagram in Figure 5.3(b) appear on opposite sides of the line diagram to either side of the central hinge *B.* This occurs because both horizontal reactions act in the same direction, so that the moments in both columns, caused solely by these reactions, will create tensile stresses on the right-hand faces and compressive stresses on the left. This can be visualized from the deflected form of the structure in Figure 5.3(c), in which the upper portion of the structure is tending to move to the right. In a rigid joint the angle between the jointed members is preserved, unlike the members meeting at the node of a pin-jointed truss. The beam and column will still be at right angles to each other in the loaded condition. The bending moments therefore cause tension on the outside of the structure and compression on the inside to the right of *B,* whilst to the left of *B,* there is compression on the outside and tension on the inside.

The maximum bending moments in the beam occur at *D* and *E,* and reduce as the central hinge *B* is approached, at which point the value is zero. This occurs because of the opposite sense of the moments caused by the vertical and horizontal reactions in each case. For example,

$$
\begin{aligned}
BM \text{ 2 metres to the right of } D &= \quad 2.H_L \quad - \quad 2.V_L \\
&\qquad \text{(clockwise)} \quad \text{(anti-clockwise)} \\
&= (2 \times 2.5) \ - \ (2 \times 1.25) \\
&= 5 \ - 2.5 \\
&= 2.5 \text{ kN-m}
\end{aligned}
$$

BM 2 metres to the left of E =
$$\begin{aligned}
&\quad 2.H_R \quad - \quad 2.V_R \\
&\text{(clockwise) \quad (anti-clockwise)} \\
&= \quad (2 \times 2.5) \quad - \quad (2 \times 1.25) \\
&= 5 \quad - 2.5 \\
&= 2.5 \text{ kN-m}
\end{aligned}$$

The whole of the bending moment diagram is linear because there are no squared quantities in the expressions.

Vertical uniformly distributed load

The analysis in the previous section has demonstrated that a three-hinged form can transmit a horizontally applied force into the foundations of a building by bringing into play the continuity at the right-angled bends. The form is therefore clearly suitable for the framework of a building in which there is no adequate arrangement of stiffening walls or service towers. If a framed building is proposed on the basis of this reasoning, it has to be accepted that the rigid joints will also behave as such under the application of the vertical dead and live loads. Using the same span and

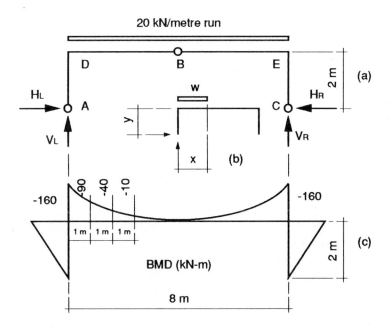

Figure 5.4   *Three-hinged portal frame – UDL*

height as in Figure 5.3, with the same notation for hinges and rigid joints, consider the effect of imposing a uniformly distributed load of 20 kN per metre run, as shown in Figure 5.4(a).

The vertical reactions at A and C will be one half of the total uniformly distributed load, i.e.

$$V_L = V_R = (½ \times 20 \times 8) = 80 \text{ kN} \uparrow$$

The frame is symmetrical this time with respect to both its geometry and its loading, so that the horizontal reactions at A and C would be expected to conform with that symmetry by both acting inwards or both acting outwards. A frame with a hinge in its span can be imagined to want to spread its feet in the absence of inward horizontal reactions. The assumed directions of H_L and H_R will therefore be towards the right and the left respectively. H_L can be found by taking moments about B from the left-hand side, thus:

$$\begin{aligned}
4.V_L &= (20 \times 4 \times 2) + 2.H_L \\
\text{(clockwise)} &\quad \text{(anti-clockwise)} \\
\therefore (4 \times 80) &= (20 \times 4 \times 2) + 2.H_L \\
\therefore 320 &= 160 + 2.H_L \\
\therefore 2.H_L &= 320 - 160 \\
\therefore 2.H_L &= 160 \\
\therefore H_L &= 80 \text{ kN}
\end{aligned}$$

Working on the left-hand side of the structure, the bending moment at D will consist, as in the previous example, solely of the moment due to H_L. This moment is the product of the force HL and the perpendicular distance between its line of action and D, i.e.

$$\begin{aligned}
\textit{BM at D} &= H_L \text{ kilonewtons} \times 2 \text{ metres} \\
&= 2.H_L \text{ kN-m} \\
&= (2 \times 80) \text{ kN-m} \\
&= 160 \text{ kN-m}
\end{aligned}$$

The bending moments between the rigid joint D and the central hinge B can be evaluated using the same sign convention and procedure as that for the simply supported beam analysed in Chapter 1, with the exception that the expression involving H_L must be included here. Unlike the simply supported beam, a bending moment, in this case 160 kN-m, already exists at the support position D. The sign convention whereby the moment due to the vertical reaction, R_L for the simply supported beam but V_L here, is positive, is retained. Under this convention, positive bending moments throughout the length of the simply supported beam resulted in tension along the whole of the underside and compression along the whole of the topside. In this example, the bending moment values at one metre intervals can be evaluated by considering forces to the left of each section, according to the general expression involving the co-ordinates x and y in Figure 5.4(b), i.e.

$$BM \text{ at } x = V_L.x - wx^2/2 - H_L.y$$

The negative sign in front of the second and third terms appears because the moments about x of the partial uniformly distributed load wx and the horizontal reaction H_L are acting in an anti-clockwise sense, whereas the moment due to the vertical reaction V_L is clockwise. The shape of the curve will be parabolic, being of the form

$$Ax^2 + Bx + C$$

where $A = -w/2$, $B = V_L$ and $C = -H_L.y$

Using the general expression and substituting the appropriate values of $V_L = 80$ kN, $H_L = 80$ kN and $w = 20$ kN/metre run,

$$\begin{aligned}
BM \text{ at 1 metre from } D &= (80 \times 1) - (20 \times 1^2/2) - (80 \times 2) \\
&= \quad 80 \quad - \quad 10 \quad - \quad 160 \ = -90 \text{ kN-m} \\
BM \text{ at 2 metres from } D &= (80 \times 2) - (20 \times 2^2/2) - (80 \times 2) \\
&= \quad 160 \quad - \quad 40 \quad - \quad 160 \ = -40 \text{ kN-m} \\
BM \text{ at 3 metres from } D &= (80 \times 3) - (20 \times 3^2/2) - (80 \times 2) \\
&= \quad 240 \quad - \quad 90 \quad - \quad 160 \ = -10 \text{ kN-m}
\end{aligned}$$

If the calculations for the reactions have been correct, the bending moment at the hinge B, four metres from D, should work out to zero, since a hinge, by definition, is a position of zero bending moment.

$$\begin{aligned}
BM \text{ at 4 metres from } D &= (80 \times 4) - (20 \times 4^2/2) - (80 \times 2) \\
&= \quad 320 \quad - \quad 160 \quad - 160 = 0,
\end{aligned}$$

as predicted.

An identical set of bending moment values will be derived by working to the right of the central hinge B. If all these results are plotted perpendicular to the line diagram of the frame, the bending moment diagram in Figure 5.4(c) will emerge. All the bending moment values are negative, implying that the bending stresses on the whole of the outside of the frame will be tensile, and that those on the inside surface will be compressive. The horizontal portion of the frame, the beam, therefore attracts bending moments of the opposite sense to those in a simply supported beam.

General expression for reactions and bending moments

If the vertical uniformly distributed load in the previous example were expressed as *w* kN per metre run, and the span and height of the three hinged frame as *L* and *h* respectively, general expressions for the horizontal reactions at the bases and for the bending moment at the rigid joints could be derived.

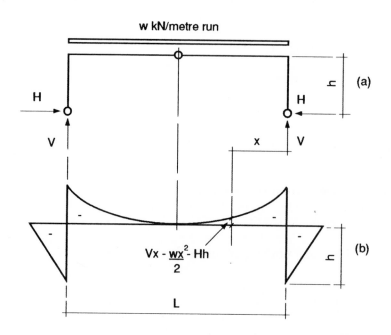

Figure 5.5   *Three-hinged portal frame – algebraic UDL*

Using the same logic as in the numerical example in the previous section, the vertical reactions in the frame in Figure 5.5(a) will be one half of the total load, i.e.

$$V_L = V_R = w \text{ kN/m} \times L/2 \text{ metres} = wL/2 \text{ kN}.$$

Since the reactions are equal, they can be simply referred to as *V*. Similarly, because $\Sigma H = 0$ and there are no external horizontal forces, the horizontal reactions will be numerically equal whilst acting in opposite directions. Both H_L and H_R can therefore be denoted by *H*. By taking moments about the central hinge of forces to the left-hand side,

$$V. \frac{L}{2} = w. \frac{L}{2} \frac{L}{4} \quad + \quad H.h$$

Substituting for V,

$$\frac{wL}{2} \quad \frac{L}{2} = \quad \frac{wL^2}{8} \quad + \quad H.h$$

$$\therefore \quad \frac{wL^2}{4} \quad = \quad \frac{wL^2}{8} \quad + \quad H.h$$

$$\therefore \quad H. \ h \quad = \quad \frac{wL^2}{4} \quad - \quad \frac{wL^2}{8}$$

$$\therefore \quad H. \ h \quad = \quad wL^2 \left(\frac{1}{4} - \frac{1}{8} \right)$$

$$\therefore \quad H. \ h \quad = \quad \frac{wL^2}{8}$$

$$\therefore \quad H. \quad = \quad \frac{wL^2}{8h} \quad kN$$

By separating h from this result, the magnitude of H can be seen to be equal to $wL^2/8$ divided by the height of the portal frame. Recalling that $wL^2/8$ is the general formula for the bending moment at mid-span for the same loading on a simply supported beam, the expression for H reveals that the magnitude of the horizontal reaction is this equivalent bending moment divided by h.

An expression for the bending moment at the rigid joints can again be derived in the same manner as in the numerical example, i.e.

$$BM \ (at \ D \ and \ E) = H.h$$

Substituting for H,

$$BM = \frac{wL^2}{8h} \times h = \frac{wL^2}{8} \quad kN\text{-}m$$

By referring to Figure 5.4(c) and comparing the shape of the bending moment diagram with that in Figure 5.5(b), the maximum bending moment on the system is $wL^2/8$. Its magnitude is the same as that for a simply supported beam with the same span and equally loaded. It occurs, however, at each end of the beam, that is at the rigid joint with the column, rather than at the mid-span position. The position of the bending moment which will determine the maximum depth of the section has, by comparing the frame to the simply supported beam, been moved from mid-span to the ends.

In the bending moment diagram in Figure 5.5(b), the negative sign has been introduced, using the outcome of the numerical results in the previous section. To prove the validity of this assumption, the algebraic bending moment value for any distance x can be written as follows:

$$\text{For } x=0 \text{ (at support), } BM = V.x - wx^2/2 - H.h$$
$$= 0 - 0 - wL^2/8 \text{ kN-m}$$

For $x \rangle 0$ and $x \langle L/2$ (between support and central hinge)
$$BM = V.x - wx^2/2 \qquad\qquad - H.h$$
(Equivalent BM for simply supported beam) $\qquad - wL^2/8 \text{ kN-m}$

Since the equivalent bending moment for a simply supported beam will, at any other position than at mid-span, be less than $wL^2/8$, the bending moment in this zone will always be negative.

At mid-span, at the central hinge, when $x = L/2$, the equivalent simply supported bending moment is equal to $wL^2/8$, so that the net bending moment at the hinge is, as expected, zero, i.e.

$$BM \ (x=L/2) = V.x - wx^2/2 \qquad - \qquad H. h$$
$$= + wL^2/8 \qquad - \qquad wL^2/8 \qquad = \qquad 0$$

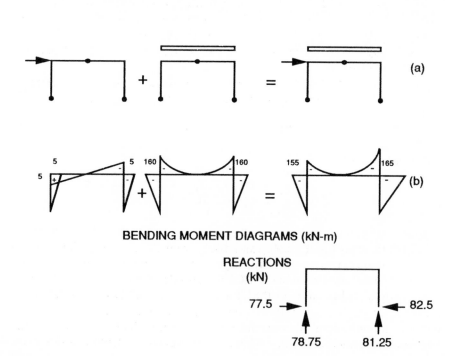

BENDING MOMENT DIAGRAMS (kN-m)

**REACTIONS
(kN)**

Figure 5.6   *Three-hinged portal frame – combined loading*

Design procedure

It can be seen by referring to the numerical examples that the effect of a lateral load is to increase the bending moment at one of the rigid eaves joints, and to reduce it at the opposite one. If the vertical and horizontal loads in those examples were applied simultaneously, as in Figure 5.6(a), the bending moment diagram for the combined load condition could be drawn by superimposing the diagram for the first condition on the diagram for the second. By doing so, the bending moment at each position on the frame is expressed graphically as the sum of the bending moments for the two separate cases.

For example, the bending moment at E from the vertical uniformly distributed load is -160 kN-m, and from the horizontal point load is -5 kN-m. By referring to the appropriate points in the text, it will be recalled that the negative sign implies tension on the outside faces of the frame, and compression on the inside faces. Thus,

$$BM \text{ at } E = (-160 -5) = -165 \text{ kN-m.}$$

At D, on the other hand, the bending moments for the separate load conditions have opposite signs, that is -160 kN-m due to the vertical load and +5 kN-m due to the horizontal load.

$$\therefore BM \text{ at } D = (-160 +5) = -155 \text{ kN-m.}$$

Figure 5.6(b), as well as showing the bending moment diagram for the combined load condition indicates the final vertical and horizontal reactions. These can also be obtained by the algebraic addition of the values for the two separate conditions, i.e.

$$
\begin{aligned}
V_L &= (80 - 1.25) &= 78.75 \text{ kN}\uparrow \\
V_R &= (80 + 1.25) &= 81.25 \text{ kN}\uparrow \\
H_L &= (80 - 2.5) &= 77.5 \text{ kN}\rightarrow \\
H_R &= (80 + 2.5) &= 82.5 \text{ kN}\leftarrow
\end{aligned}
$$

A quick arithmetical check can be made to verify that the Newtonian equilibrium state is still maintained, i.e.

$$
\begin{aligned}
\text{Total applied vertical force} &= (20 \times 8) &= 160 \text{ kN}\downarrow \\
V_L + V_R = (78.75 + 81.25) &= 160 \text{ kN}\uparrow & \therefore \Sigma V = 0.
\end{aligned}
$$

$$
\begin{aligned}
\text{Total applied horizontal force} &= 5 \text{ kN}\rightarrow \\
H_L + H_R = (77.5 - 82.5) &= 5 \text{ kN}\leftarrow & \therefore \Sigma H = 0.
\end{aligned}
$$

The effect of the horizontal point load on the symmetrical three-hinged portal frame is therefore to transfer some of the vertical load from the left-hand to the right-hand support. It also causes an increase in the horizontal reaction at the right-hand support, with a corresponding decrease at the left.

If the 5 kN horizontal point load were applied at E and acted towards the left instead of to the right, the reactions obtained above would have the same numerical values, except that the conditions at the two supports would be reversed, as would the bending moments at D and E.

Where the bending moments arising out of horizontal wind forces on a portal frame are less than about a quarter of those due to the dead and live loads, the design codes usually allow them to be ignored. This is because the overstressing of the materials is not permanent, and therefore creates only a relatively small, temporary encroachment on the factor of safety on which the design is based. Wind forces would only tend to govern the design where the frame is tall with a relatively small span. The magnitude of the bending moments at D and E due to the horizontal wind force would then become greater as compared with those due to the vertical dead and live loads.

Pitched portal frames

As a structural form for a single storey space, the rectangular portal frame is less familiar than its pitched roof derivative. The behaviour of this form can be more clearly appreciated by looking on it as a deviation from the rectangular form analysed in the preceding sections. It becomes pitched rather than rectangular simply by moving the central hinge upwards, as shown in Figure 5.7, with the consequent lengthening of the original horizontal member, which is now split into two parts.

The maximum bending moment in a pitched portal frame subject to a vertical uniformly distributed load will still nearly always occur at the junction between the column and the beam, more accurately described in this case as a sloping rafter. The magnitude of that bending moment will, however, be less than that occurring at the beam to column junction, i.e. $wL^2/8$, in the rectangular portal frame carrying an identical load. If, for example, the height of the column were eight tenths of the height h of the central hinge, as shown in Figure 5.7(a), the calculation for the maximum bending moment would be as follows:

$$H = \frac{wL^2}{8h} \text{ kN, as for the rectangular frame.}$$

If calculated from first principles, the equation yielding this formula for H would be identical to that for a rectangular portal frame of height h,

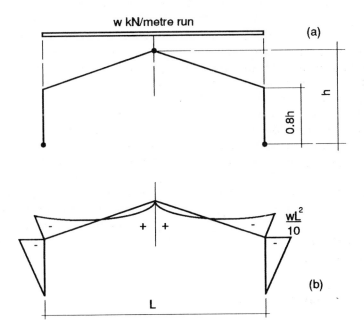

Figure 5.7 *Pitched portal frame – UDL*

since the only vertical dimension involved is *h*. The height of the column is not relevant at this stage. The two heights in the rectangular portal frame just happen to be the same.

The bending moment at the column to rafter junction is then simply the product of *H* and the height of the column, i.e.

$$BM = H \times 0.8h$$
$$= \frac{wL^2}{8h} \times 0.8h$$
$$= \frac{wL^2}{10} \text{ kN-m}$$

If the height of the column were 0.5*h*, the pitch being much steeper, the bending moment at the column to rafter junction would have been correspondingly smaller, i.e.

$$BM = H \times 0.5h$$
$$= \frac{wL^2}{8h} \times 0.5h = \frac{wL^2}{16} \text{ kN-m}$$

Figure 5.7(b) shows the shape of the bending moment diagram for a pitched portal frame, in which the bending moment changes from a negative to a positive vector at a certain point along each rafter. This point, at which the bending moment is zero, is known as the point of contraflexure, the prefix 'contra' implying a change in the nature of bending, or flexure, in this case from hogging to sagging. In most pitched portal frames encountered in practice, the maximum positive, or sagging, moment at the two positions near to the central hinge will be much less than the maximum negative moment at the column to rafter junction. For the positive rafter moment to be greater than the negative moment at this junction, short columns would need to be combined with a steep pitch. There is an extreme condition where the pitch of the rafters starts at the support positions, the columns being reduced to a height of zero and therefore ceasing to exist. Here, there will be no negative bending moments, but only positive moments with maximum values half way along each rafter. The disappearance of columns deprives such a structural form of its status as a portal frame, converting it into a pair of inclined beams.

The portal frame in architecture

Grid systems

The reinforced concrete framed building is arguably the most commonly occurring example of the application of the portal frame principle to the enclosure of space. Yet it is in such buildings that this structural form is nearly always concealed, externally by the curtain walls and internally by suspended ceilings. In those cases where the form is unashamedly expressed, the structure could well have been the primary determinant of the whole design concept, forcing the architect into a deliberate suppression of the non-load-bearing elements of the building. The only way, for example, in which a single storey three-hinged form could be openly expressed on the outside would be to keep the enclosures inside the columns, and to suspend the roof slab from the underside of the beams. Conversely, internal expression of the frame would involve putting the walls and roof on the outside of the beams and columns. A totally glazed building would allow the portal frames to be seen both from inside and from outside the building, but the design would hardly be found to be energy-saving or free from the effects of solar glare.

The main condition implicit in the adoption of a framed structure on a regular grid as a solution to an architectural brief is that all the activities which are to take place within the building can be related to the geometry of that grid. This condition would be satisfied if a building structured with frames at three-metre centres were to be divided into office or shop units three, six or nine metres long. Such a framed system imposes its own discipline on the building. If the internal spaces conformed to an irregular geometry, a framed system could still be used, provided that the structural grid were compatible with the planning grid. In such cases, the merits of a framed structure should be carefully weighed against a load-bearing wall system, which would allow the usable spaces to define the structural grid. It is rarely advisable to impose a grid of equally spaced portal frames on a building of this nature. Such a decision could result in awkward, even impossible junctions of transverse partitions with windows or curtain wall panels set between the frames.

At the other extreme, a portal framed system is one of many possible solutions to the problem of covering a large span. The main determinant of the spacing of the frames here will be the available dimensions of the roof and wall elements which must fit between the frames. On the other hand there may be a preferred elevational geometry if the architect is an advocate of the strict rules of classical proportion.

There is no such thing as an ideal spacing of columns in a regular grid system. Even if a spacing is determined from the constraints mentioned above, this may not be the ideal from the point of view of structural economy. Portal frames with a close spacing will result in small roof, floor and possibly wall spans between them, and therefore secondary structural elements with small bending moments and minimal depths. A greater spacing will require a smaller number of portal frames, but deeper secondary elements. These widely spaced frames, because they are carrying greater loads than the closely spaced frames, will also have greater structural depths, and will consequently be heavier. This, in turn, could affect the means and possibility of transportation if the portal frames are fabricated away from the site.

Frank Lloyd Wright and the portal frame

The buildings of Frank Lloyd Wright (1868–1959) have exerted a seminal influence on twentieth century architects for a variety of reasons, the most understated of which is probably his awareness of the expressive power of structure. One of the most subtle relationships between the principle of the three-hinged portal frame and the quality of an architectural space occurs in the structure of Wright's Administration Building of the Johnson Wax Corporation in Racine, Wisconsin, USA, completed in 1939 (Plate 9). The reinforced concrete column and roof modules have frequently attracted a visual comparison with mushrooms. A single mushroom, however, has its own independent botanical and structural existence, until picked and eaten, not relying in any way on its neighbours for support. Its column has a fixed base.

Plate 9 *Johnson Wax Building by Frank Lloyd Wright. Source: Ezra Stoller.*

Wright's columns, on the other hand, depend for their stability on the mutual contact between each circular slab and its four adjacent units. Thus, a section taken through any row of columns reveals a structural form which tapers, in response to decreasing bending moments, towards three hinges, two at the column bases and one at the intersection of the circular slabs. It is this latter junction which corresponds to the central hinge in a planar three-hinged frame. Since the same structural form would appear if the section were also drawn at right angles to the first one, the whole arrangement appears as an orthogonal grid of intersecting three-hinged frames. No attempt is made to conceal the structure internally. The fusion of architecture, interior design and a minimal structural form has been achieved by conveying the essence of a simple structural principle in an unfamiliar manner. A form which usually appears in two dimensions as a series of plane frames supporting a secondary structure spanning at right angles to it has acquired a third dimension.

Longitudinal stability

Unlike Wright's three-dimensional structure, the majority of buildings consisting of one-way spanning frames do not have any inherent stability in the third dimension, that is parallel to the length of the building. Horizontal wind forces in the plane of the frames applied to the sides of a building can be collected as point loads at the eaves and transmitted to the foundations by virtue of the rigidity of the beam to column junctions in each portal frame. Applied to the ends of a building, such forces would act at right angles to the frames, rendering them liable to rotation about their bases and eventual collapse, especially if their bases are pinned. Fixed bases, if the horizontal forces acting on the structure are shared between all of the columns, will offer some resistance as vertical tree structures. One disadvantage of this solution is that large horizontal deflections at their tops may cause distortion of and damage to wall and roof claddings.

The problem can be overcome by providing an edge beam with rigid connections to the tops of the columns, thus creating a multi-bay portal frame on each side of the building. This is easily achieved if the structure is to be of *in situ* reinforced concrete construction, in which it is relatively easy to design and construct a two-way moment connection by careful attention to the detailing of the reinforcement. If the portal frames are precast, it may prove difficult to connect the edge beams to them as a rigid joint. Structural steelwork also lends itself to the introduction of rigidly connected edge beams when 'I' section universal columns are used, although this does impose bending moments on the weaker axis of the column. A laminated timber portal frame, whether the rigid moment connection is achieved by bolting through the separate beam and column components or by an organic continuation of the laminations from the beam into the column, clearly does not lend itself to continuity with an orthogonal edge beam.

Unless the structure is of *in situ* reinforced concrete, by far the simplest way of achieving longitudinal stability is by means of diagonal bracing.

Depending on the magnitude of the wind forces and the desired elevational treatment of the building, the bracing can be provided in every space between adjacent columns, or in a few positions only. It is possible, for example, to brace only the two end bays, or only the central bay. The more bays that are braced, the smaller can be the cross-sectional area of the diagonals for a given total horizontal force.

Consider a single bay whose share of the total horizontal force on the entire structure is 5 kN. If this is applied from the left, as shown in Figure 5.8(a), the edge beam will transfer the 5 kN as a strut to the top of the diagonal, in which the force can be evaluated using the triangulation principles described in Chapters 1 and 4.

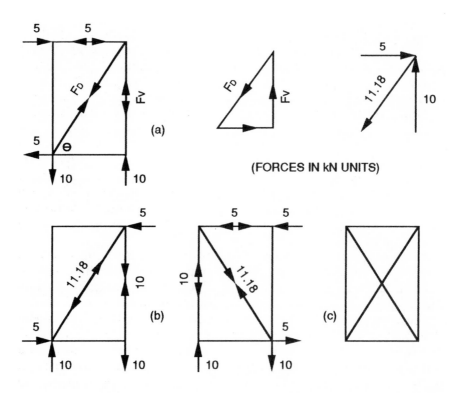

Figure 5.8 *Longitudinal bracing system*

If the columns are spaced at 2 metre centres and their height is 4 metres, the angle θ between the diagonal member and the horizontal will be such that

$$\tan \theta = 4/2 = 2$$
$$\therefore \quad \theta = 63.4°$$

Denoting the forces in the column and the diagonal as F_V and F_D respectively, the vertical and horizontal components of the three forces meeting at the top of the diagonal can each be summated to zero, i.e.

$$\Sigma V = 0, \therefore F_V = F_D.\sin 63.4° = 0.894F_D$$
$$\Sigma H = 0, \therefore 5 = F_D.\cos 63.4° = 0.447F_D$$

From the second expression, $F_D = 5/0.447 = 11.18$ kN. Substituting this value in the first expression,

$$F_V = 0.894 \times 11.18 = 10 \text{ kN}$$

The triangle of forces shows that the vector arrow in the diagonal is pulling away from the node point. This member is therefore subjected to a tensile force of 11.18 kN. The vertical force of 10 kN in the column is pushing towards the node point and is therefore compressive. This would have to be compared with the axial forces in the column arising from dead and live loads to establish whether the additional compressive stress will influence its size. The edge beam which carries the 5 kN horizontal force to this node point would also need to be checked as a strut in addition to its function as a lightly loaded beam.

Also shown in Figure 5.8(a) are the additional loads imposed on the column bases by the vertical component of F_D and the force F_V, both equal to 10 kN but acting in opposite directions. Their opposition is a logical outcome of the analysis because no vertical forces have been applied to this structural sub-system, in which the external forces must all be in equilibrium, i.e.

$$\Sigma V = 10 \text{ kN}{\downarrow} + 10 \text{ kN}{\uparrow} = 0$$

Similarly, the horizontal component of F_D is equal to 5 kN and will be resisted at the left-hand base, so that for external forces

$$\Sigma H = 5 \text{ kN}{\rightarrow} + 5 \text{ kN}{\leftarrow} = 0$$

If the wind were blowing from the opposite direction, the 5 kN force would be acting from right to left, as shown in Figure 5.8(b). If the diagonal member were oriented as in 5.8(a), the triangulation at its top would show F_D, whilst still subjected to a force of 11.18 kN, to be in compression. The triangle of forces has the same geometry, but the vector arrows face in the opposite direction. Since a strut, for the same force and effective length, will need to have a greater sectional area than a tie, it makes more sense to introduce a second diagonal member. This is also shown in Figure 5.8(b). In this arrangement the force F_D in the diagonal is again tensile. Figure 5.8(c) shows the resulting cross-braced arrangement, in which both diagonals respond as tensile members according to the direction in which the wind is blowing.

The results of the triangulation which eventually yielded the magnitude and direction of the two vertical reactions can be verified by taking moments about the ground of the original 5 kN horizontal force. If the reaction at the right hand column in Figure 5.8(a) is denoted as R, then, by taking moments about the left-hand column,

$$(5 \times 4) = (R \times 2)$$
$$\text{(clockwise)} \quad \text{(anti-clockwise)}$$
$$\therefore 20 = 2R$$
$$\therefore R = 20/2 = 10 \text{ kN}$$

The horizontal distance of 2 metres, the column spacing, can be regarded as the structural depth of this sub-system.

Lest the impression has been conveyed that horizontal forces arise only from the wind, it is important to remember that any plane structural element such as a portal frame is, in isolation, in a condition of unstable equilibrium. It will tend to rotate about an axis joining its column bases, away from its equilibrium position rather than towards it. The horizontal force vector used in the foregoing calculations could therefore equally apply to a frame's inherent instability perpendicular to its span in a theoretically windless location. A coherent system for channelling longitudinal forces to the ground is vital for a building's safety both in its completed state and during all stages of construction.

The fallacy of the single-hinged base

Buildings in which portal frames are openly expressed usually convey a sense of structural harmony as well as of spatial order. A visible base hinge, appearing as one of a pair in a two or a three-hinged frame, does not portend imminent collapse. The steel frame in Plate 10, with its single-hinged column and off-centre location, would undoubtedly start to rotate about its base if left to its own devices. It is prevented from doing so by its connection, through the secondary structural elements at right angles to it, to the internal masonry towers almost concealed from view at the end furthest from the column. Each steel frame, one on either side of the building, is deliberately emphasized. The external walls and the masonry access towers appear in receding vertical planes.

Plate 10 *Rome-Florence Motorway Restaurant*

Plate 11 *Base Hinge of Motorway Restaurant*

The building, a service area restaurant on the Florence to Rome motorway, can create a sense of unease enhanced by the clearly stated function of its base hinge in Plate 11. It is unusual to see a hinge in the context of the entire frame with which it is associated. One of the earliest buildings to express a hinge so forcibly was the AEG Turbine Factory in Berlin (1909–10), designed by Peter Behrens (1868–1940). Structured by a series of parallel three-hinged steel portal frames, each hinge conveys the message that it is part of a structural form possessing another hinge on the opposite elevation of the building. Whilst there is another hinge on the opposite elevation of the motorway restaurant, it is not associated with the behaviour of the visible portal frame. Each single-hinged frame is spanning in the unexpected longitudinal direction.

Even if this idiosyncrasy is regarded as misleading and structurally dishonest, its form is still expressed through the rich vocabulary of the portal frame.

Arch forms

Different structural forms may at first sight be perceived as expressions of separate, unconnected modes of physical behaviour. It is convenient, after all, for the human brain to classify apparently unrelated phenomena according to some preconceived ideas of like and unlike. Applied to the study of structure, however, an approach as simplistic as this would diminish rather than enhance a deeper understanding of the underlying principles.

An appreciation, on the other hand, of the similarity that exists between outwardly unconnected structural configurations can lead the architect towards an intellectual habit of making informed choices. Any selection between alternatives can only be effective if the precise nature of those alternatives is clearly understood. The next two sections will be concerned with the similarities that exist between the arch and the portal frame.

Three-hinge configurations

Stability of the triangle

It was shown earlier that the notion of connecting a beam to a column by a moment connection can yield forms which are stable under the action of imposed vertical and horizontal forces. The simplest of such forms, the rectangular three-hinged portal frame, was then analysed using the laws of static equilibrium. By joining the hinge positions of this frame, as shown in Figure 5.9(a), a triangle is formed. This diagram suggests that the inherent stability of this form has something to do with the stability of a triangle.

By juxtaposing the triangle with the portal frame, as in Figure 5.9(b), it may be recalled from Chapter 4 how important was the geometry of the triangle to the design of pin-jointed trusses. In both the triangle and the portal frame, failure can only occur by breaking the parts between the hinges. Neither form can be distorted in the same way as can the four-hinged frame in Figure 5.9(c), in which the original square is easily distorted into a rhombus.

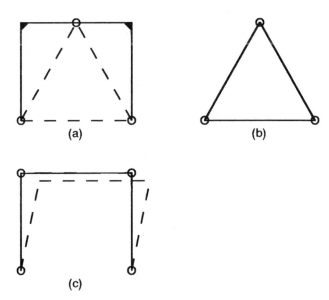

(a)

(b)

(c)

Figure 5.9   *Stability of triangulated form*

This can be simply demonstrated by making a balsa or card model of each of the forms, connecting each pair of adjacent sides with real pins. A pin or a hinge will join the members together, whilst allowing them to rotate with respect to each other. In contrast, the beam to column junction, often referred to as the eaves junction, in the three-hinged portal frame is fixed, in that a bending moment can be transferred across the joint.

Infinite number of three-hinged forms

Now that the principle of a three-hinged form has been established, there exists a clear possibility that there is more than one possible shape for the parts between the hinges. Figure 5.9 shows two such shapes: a straight line and an 'L' shape. Provided, therefore, that these configurations are strong enough to resist the applied bending moments at all points between the hinges, in that their structural depth is always compatible with the magnitude of these moments, there are not just two, three or even four, but an infinite number of possible shapes.

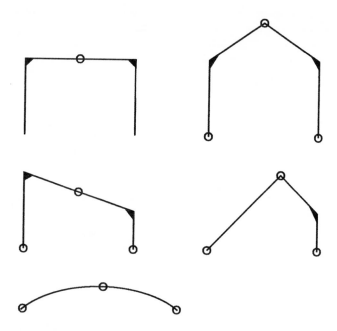

Figure 5.10   *Three-hinged forms*

Of the five shapes shown in Figure 5.10, the top two are familiar, the middle two less so, and the lowest is noticeably curved. But all five belong to the same family of structures, even if only one or two are recognizable as part of the vocabulary of the built form. The clue to the link between the portal frame and the arch, lies perhaps in the appearance of the curved shape.

There are two concepts vital to the understanding of arch and frame behaviour. The first is that of perfection in structural form. The second involves the similarity in behaviour between sections in structural forms which are constrained to develop resistance moments, whether they be beams, portal frames or arches. The design methods outlined in the preceding chapters can, if the principles are fully understood, be applied equally well to the design of a three-hinged arch as to the first two more familiar forms.

The arch in architecture

The Romanesque precedent

The achievements of the Romans in conceiving and building aqueducts such as the Pont du Gard at Nimes in France and that at Segovia in

Spain in the first century AD bear sufficient witness to the capacity of the semi-circular arch to support its load over a long period of time. These structures, however, despite their total domination of their respective landscapes, are not space-enclosing. They belong to the world of engineering rather than to that of architecture. To appreciate the internal spatial qualities engendered by the semi-circle, we have to go forward by about a thousand years to the full flowering of the Romanesque style, to the time of the building of the first Gothic cathedrals.

'With the Cathedral of Durham we reach the incomparable masterpiece of Romanesque architecture not only in England but anywhere.'

This was Alec Clifton-Taylor's judgement (Clifton-Taylor, 1967, p.42). It was not only in the arches between the massive circular piers beneath the triforium that expressed the harmony of a repeating geometric form. More eloquent in pure structural vocabulary is the chevron-decorated semi-circular form of the arches in the Galilee Chapel. The columns owe their slender proportions to the correctness of the builder's judgement that the horizontal reaction required for the stability of an arch can be exactly balanced by that required for the stability of an adjacent equally loaded arch of the same span. Since the ratio of dead load to live load is very high in this case, the horizontal reactions at a column supporting any two arches will invariably be almost equal. The column therefore has only to resist a vertical load, and is relieved of the need to act also as a buttress.

Durham was completed early in the twelfth century. During the next four centuries the manner of vaulting developed towards forms later identified as more valid expressions of the laws of structural mechanics. One of the main features of the Gothic style is the pointed arch, the curved portions of which are, at least visually, nearly parabolic. The principal objective for the master masons was to achieve great height in the nave, transept and choir, rather than to engineer wide spans. The aesthetics of the interiors of some cathedrals have been criticized for not having small enough spans. Clifton-Taylor asserts that 'York nave is yet another English Gothic building which is too broad for its height' (Clifton-Taylor, 1967 p.168). York Minster has the highest nave of the English cathedrals, if Westminster Abbey is excluded by its present status as an abbey church. A smaller span would have given greater emphasis to its 28 metre height. This is not to assert that the problem of span was in any way trivial. The span of the nave at York is nearly 15 metres – a considerable achievement in stone with its minimal tensile strength.

Funicular forms

The search for perfection in a spanning structural element is synonymous with the quest for a form in which all bending moments are eliminated.

Comparing a simply supported beam (Figures 1.8 and 1.12) with a rectangular portal frame (Figure 5.4), each loaded with a UDL, the first exhibits positive bending throughout and the latter negative bending. It can therefore be inferred that there must exist a form somewhere between these two geometries in which the bending moment at all positions will be zero. Using the expression for the bending moment in a portal frame derived earlier in this chapter, i.e.

$$BM = \text{equivalent } BM \text{ for simply supported beams} - H.h$$

a zero bending moment occurs when the term $H.h$ is numerically equal to the bending moment for the 'beam condition'. Since H is a constant for any given loading, this will occur at every point in the system when h traces out the shape of the bending moment diagram for the 'beam condition', that is a parabola.

The parabola can therefore be described as the funicular form for a uniformly distributed load. It may reasonably be assumed that every other pattern of loading has its own distinct funicular form. A similar relationship to that which exists between a simply supported uniformly distributed load and its bending moment diagram can be shown to exist between, for example, a central point load and its bending moment diagram.

By a similar chain of reasoning to that pursued for the uniform load condition, what is sought is a structural form in which the bending moment at any position for the 'beam condition' is cancelled out by the moment due to the horizontal reaction H. The form which emerges is, predictably, a triangle. It is hardly surprising that the funicular form, when its ends are joined up to the original line diagram of the beam, completes a triangle. The simplest truss explored in Chapter 4 was, after all, an isosceles triangle, in which the two equal, inclined sides of the triangle were shown to be the compression members. The word 'funicular', incidentally, is the adjective derived from the noun 'funicle', which is defined in the dictionary as a small cord or a fibre. The implication for structure is that of an element having minimal thickness, and therefore negligible resistance to bending. It will be demonstrated in the next chapter that a tensile funicular form is always in a more stable condition than a compressive one.

The Gothic vault

As far as can be ascertained, the master masons responsible for the design and construction of medieval churches and cathedrals did not know how to draw bending moment diagrams, nor did they spend what little spare time they had discussing funicular forms. Miraculously, they did manage to devise, either collectively or by the inspired thinking of a few individuals, a form of vaulting almost identical in shape to the combined bending moment diagram for a uniformly distributed load and a central point load. The loading and the bending moment diagram for this condition are shown in Figure 5.11. It can easily be deduced from the arguments relating to each of these conditions in their separate states that this bending moment diagram is the funicular form for the combined loads.

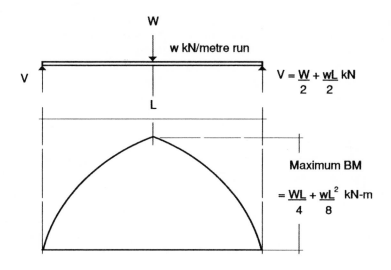

Figure 5.11 *Combined UDL and point load*

The expression for the curves to each side of the centre line is

$$Vx - wx^2/2$$

and is therefore of the form

$$Ax^2 + Bx + C, \text{ where } A = -w/2, B = V \text{ and } C = 0$$

so that they are parabolic. What has happened is that each half of the parabola has been tilted outwards, producing a cusp at mid-span. This effect is caused by the magnitude of V being increased by half the value of the point load over the reaction from just the uniformly distributed load. Now it may be supposed that a Gothic vault would impose an approximately uniformly distributed load upon itself. This is certainly the impression gained from inside a vaulted space, from which none of the external loads or support systems is visible. But the function of a stone vault is to act as a ceiling. Stone is a non-combustible material, so that the building would possess a greater measure of resistance to an internal fire than if the timber roof structure were exposed on the underside.

The pitched roof form over the vault adopted for ecclesiastical buildings did lend itself to the triangulation methods discussed in Chapter 4. Indeed, the possibility of spanning the entire width of the nave may well have occurred to the builders. But with a potential support position at mid-span, coinciding with the highest point of the stone vault, there was an immediate option to divert about half the load from the pitched roof to the vault as a central point load.

Transfer of horizontal reactions

If the form of the vault arose out of the inspired application of precedents in nature and in some earlier constructions by man, the means of providing reactions at its support points, or springings, appears in retrospect even more ingenious. The idea of the flying buttress acknowledges the necessity of moving the horizontal force at the base of the vault to a position where it can be resisted more easily. The reaction H at the springing, remembering Newton's Third Law, becomes an action as far as the support position is concerned. The same applies to the vertical reaction V. The vertical action can be absorbed provided that the wall is not too slender.

A horizontal force on a structural element, however, demands that this element behave as a vertical cantilever, in which the tensile stresses may be partially or totally offset by compressive stresses from axial loads. If the columns beneath the springings of the vaults were to be kept as thin as possible, as was the aim in the later Perpendicular style, their resistance to bending moments arising out of an additional role as a vertical cantilever would be minimal. By making the piers on the other side of the aisles sufficiently deep in their plane of bending, both the section modulus and the self weight became of an order compatible with resistance to horizontal forces. The flying buttress, acting as a strut, transfers the horizontal action across the aisle to the top of this pier, which can now perform as both a vertical cantilever and a column. Plate 12 shows this arrangement outside the nave of Canterbury Cathedral. The flying buttress itself has an arched soffit, the reason being that a stone buttress with a rectangular profile would be subject to quite considerable tensile stresses along its underside owing to bending moments generated by its own weight. The stones would probably fall out.

Where an ecclesiastical building has no aisles, there is no alternative but to make the columns in the enclosure of the nave deep enough to resist the horizontal forces from the vaults. A stunning example of this

Plate 12   *Canterbury Cathedral*

occurs in King's College Chapel in Cambridge, completed early in the sixteenth century. The depth of the piers increases towards the ground as a series of steps, keeping pace with the bending moment diagram which increases to a maximum at ground level. At this level, the piers are not externally visible owing to the presence of the side chapels. The prominent pinnacles erected on top of each pier clearly make some contribution to keeping the resultant vertical load within the middle third of each pier, thus avoiding tensile stresses on their internal faces.

It is often said of a system of vaults, flying buttresses and external piers that the weight of the vault is lifted clear of the columns flanking the nave. This would only be the case where the buttress is in perfect alignment with the tangent to the curve at the base of the vault, so that it provides the resultant reaction to the resultant action from the vault. The resultant reaction R, from the principle of composition and resolution of forces, would be related to V and H by the relationship:

$$R = \sqrt{V^2 + H^2}.$$

Where this alignment occurs, the vertical and horizontal reactions exist as components contained within the resultant reaction in the buttress. No vertical force is transmitted from the vault into the column beneath. Where the flying buttress is not aligned with the springing to the vault, the construction of a triangle of forces at this point would reveal that the vertical load from the vault is shared between the flying buttress and the nave column. The proportion of the load going into the column depends on the angle at which the buttress lies in relation to the vault. In the extreme condition when the buttress is horizontal, the entire weight of the vault is supported by the column, with the flying buttress providing only the horizontal reaction.

Plate 13   *Venice Fish Market*

The appearance of masonry arches

Possibly the most important aspect of arch behaviour for an architect is the relative proportioning of a masonry facade according to the magnitude and direction of the required reactions. Plate 13 shows a corner of the Venice Fish Market (1907). The size of the piers at the end of each elevation relative to the circular columns acknowledges the unbalanced horizontal thrusts from the end arches. The circular columns receive horizontal thrusts from each side. These tend to cancel each other out, leaving the columns to resist vertical forces only. The end piers must act as buttresses, which must provide horizontal reactions to ensure the stability of the arched elevations. The appearance of the elevations is made aesthetically satisfying by the contrast in scale arising out of structural necessity.

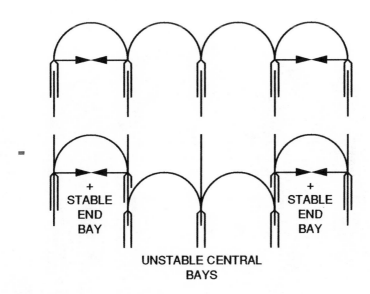

Figure 5.12 *Internal tying of arches*

A frequent sight in masonry arches in historic buildings is the horizontal tie member joining the abutments. This modification would usually have appeared some time after the building was constructed, presumably when it was noticed that the springings of the arch had started to spread. The intuitive decision to tie the bases together was, although probably not considered as such at the time, a removal of the horizontal reaction *H* from outside the arch to the space between the arch. The external force had become an internal force, although both acted in the same direction. As vectors, sharing the same magnitude and direction, they were identical. Where there is a line of arches along a facade, it is of little use to tie only the end bays when the corner columns begin to move outwards. The structure can only be stabilized by tying every arch, as in Plate 14. If only the end bays were tied, each adjacent bay would become unstable, because the ties in the outermost arches remove the balancing horizontal force from the shared supports. This condition is shown in Figure 5.12. The penultimate arches will thrust against the shared piers, which in their changed condition are in the same unstable state in which the end piers originally found themselves.

Plate 14   *Doge's Palace, Venice*

An external corner pier is in a potentially unstable condition from the outset because it is being subjected to horizontal thrusts H in two directions at right angles. If the facades are symmetrical, this causes a resultant thrust at an angle of 45° to each axis, requiring a reaction in this direction of $1.414H$. The acknowledgement of and provision for this increased force enriches the architectural quality of a structure, as in Plate 14. In contrast, the condition for a corner pier within a courtyard is in a more favourable position, as in Plate 15. This member receives vertical and horizontal forces from four orthogonal directions. Each linear pair of horizontal forces is in equilibrium, again assuming symmetrical loading. The corner pier in this case is proportioned identically to the remaining piers, reflecting an identical need to resist only vertical forces. A more massive corner pier would detract from rather than enhance the sympathy between architecture and structural form.

Plate 15   *Basilica, Assisi*

The fallacy of the half arch

A postscript to an exposition of arch behaviour concerns a fallacy that can easily arise at the sketch stage of a scheme. Figure 5.13 shows one half of an arch under a uniformly distributed load. In a complete arch, the horizontal reaction travels from one support as a constant vector to the other support. In the imagined half arch, it cannot simply disappear. There must be a horizontal reaction at the termination of this member at what would have been the crown of the complete arch.

The only rational structural completion to this elevation would be the provision of a vertical cantilever strong enough to receive a horizontal point load at its top, coinciding with the crown of the arch. The form of this vertical cantilever would be generated by the shape of the bending moment diagram, leading to a rather whimsical elevational treatment.

An apparent 'half arch' without any such end restraint could only function as a beam curved in elevation which, unless in a cantilevered mode, would still need a vertical support at its end. The proportions of this curved beam would be conditioned by bending in the actual depth of the element rather than by the true structural depth of the arch.

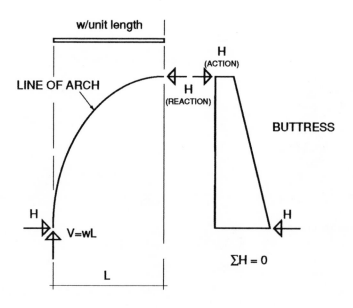

Figure 5.13 *Buttressed half arch*

Chapter 6

SUSPENSION AND CABLE STAYED SYSTEMS

Suspension structures

Chapter 5 introduced the arch as a variant of the portal frame by comparing the ways in which these forms respond to external forces. The suspension cable can also be understood in relation to the arch, but in

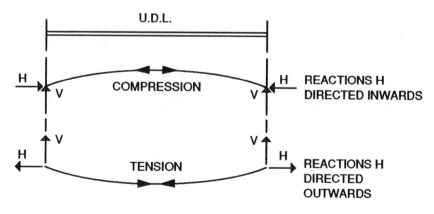

Figure 6.1 *Arch-cable analogy*

this case as an opposite rather than as a variant. At its simplest level, the perfect arch form, the parabola, supporting a uniformly distributed load projected horizontally on the span will, if turned upside down, become a parabolic suspension cable. Figure 6.1 illustrates this comparison. The compressive form changes to a tensile form. The vector arrows representing the horizontal reactions, meanwhile, have changed direction.

For a clear understanding of what the two forms have in common and of their essential differences, it is first necessary to re-examine the nature of tension and compression as they appear in curved, rather than linear forms.

Linear tensile forms

Figure 6.2(a) shows the simplest form of tensile structure, in which a single weight is suspended from the mid-point of a length of rope. The magnitude of the weight, and the scale of the structure, are not crucial to the logic to be pursued, which will involve only vector arrows and algebraic notation. The vector W represents the weight and the vector T the tensile force in the cable.

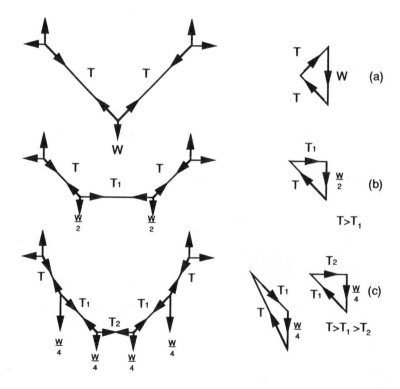

Figure 6.2   *Cable forces under point loading*

By sketching the triangles of forces at the load position and at the supports, the two parts of the rope, as would be expected from everyday experience, reveal themselves as ties.

If the weight were now split into two equal portions, the arrangement would be as shown in Figure 6.2(b). The tensile force in the central portion of the rope can be labelled T_1, and the force in the end portions of the rope T, as before. The force triangles at the load positions can again be sketched, using the vector $W/2$ for half of the weight as the basis of the triangles, which will be congruent. The force triangle for the left-hand half of the structure is shown. It is a right-angled triangle, in which the vector T, being the hypoteneuse, is clearly greater than the vector T_1.

Taking the process a stage further, if W is split into four equal portions, as in Figure 6.2(c), four force triangles involving the load positions can be drawn. T, T_1, and T_2 now represent the tensile forces in successive straight portions of the rope, working inwards from the supports. The force triangles for the two left-hand load positions are shown, in which it can be seen that the following relationship applies: $T > T_1 > T_2$.

This process could be carried on indefinitely, dividing both the original weight W and the rope into smaller and smaller units. This leads to the limiting condition at which the weight becomes a uniformly distributed load applied to a curved cable form.

However the cable is visualized, whether as a smooth parabolic curve or as a very large number of very small straight lines, it becomes clear that the tensile force has its greatest value at the ends, and its smallest value at mid-span, or where the sag is at its maximum. In the straight line analogy, the force vector T nearest to the end of the cable is always greater in magnitude than the vector adjacent to it. If the force in the small straight line element of the cable at mid-span, which will be horizontal, is given the symbol T_M, the relationship between adjacent tensile force vectors will be

$$T > T_1 > T_2 > T_3 > T_4 > T_5 > \ldots\ldots T_M$$

The physical reality of this expression is most clearly visualized in the structural form of a cable carrying several individual cars, a system familiar in skiing resorts. Figure 6.2(c) models this condition. The first impression of an occupant of such a car may well be that the cable is most highly stressed at mid-span. From the theory outlined above, the opposite is true – the greatest tension in the cable is at the ends.

The arch cable analogy

It is when the parabolic form of the cable is considered that it becomes apparent that it has some features in common with the arch. Referring again to Figure 6.1, it emerges that both the arch and the cable are funicular forms for the uniformly distributed load, because they are both parabolas. The main difference is that the arch is in compression, and the cable is in tension. There are no bending moments anywhere on either structure.

A cable can, however remain without bending moments even though the loading pattern changes. Each distinct loading system generates its own funicular form, three of which are those in Figure 6.2. The cable can easily change its shape in response to changes in loading, because each straight line element can behave as an individual tie. Its condition is such that in its stressed state, it is simply a longer straight line element than it was in its unstressed state. In common with any tensile element, its geometric identity as a straight line is enhanced rather than diminished when the stress is increased, provided that it does not reach the point of failure. A stretched elastic band models this behaviour pattern.

An arch, on the other hand, will be subjected to bending moments as soon as the loading on it deviates to the slightest extent from that for which its shape is the funicular form. If a system of point loads were to be supported by a compressive structure, a funicular form of the type shown in Figure 6.3 could be generated. But such a form can exist only as an idealized condition, because the smallest alteration to the magnitude or position of any of the loads will compel the arch to alter its geometry. It will move away from its equilibrium state and eventually fail, as will any arch containing more than three hinge positions. A tensile cable, on the other hand, will move towards its equilibrium state as its geometry changes.

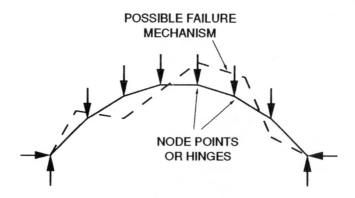

POSSIBLE FAILURE MECHANISM

NODE POINTS OR HINGES

Figure 6.3   *Instability of compressive linear form*

There have been cases where a compressive funicular form has been sustained by the action of a secondary, stiffening member whose flexural strength has prevented the displacement of any of the node points. A

particularly graphic example of this type of behaviour occurs in the
Schwandbach Bridge in Switzerland (1933), designed by Robert Maillart
(1872–1940). The thin load-bearing linear arch is restrained from
changing shape by the stiffening effect of the much deeper beam element
at the level of the roadway. The arch therefore both supports and receives
restraint from the road deck.

The washing line analogy

The reversal of the direction of the vector arrows from compression to
tension means that at the support positions of the cables, the horizontal
reaction vectors point outwards rather than inwards. This is proved in
Figure 6.4(a) by drawing the force triangles at the cable supports, and
indicating the actions produced on the supporting columns. These try to
bend inwards, such movement being resisted by their behaviour as
vertical cantilevers. The bases to these columns would need to be fixed to
the ground rather than pinned.

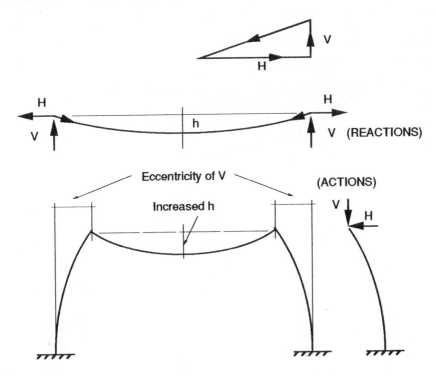

Figure 6.4 *Cable on slender supports*

The impression conveyed is that of a line laden with saturated washing, in which heroic deeds are frequently asked of hopelessly thin posts. Structures have nevertheless been built where the horizontal actions from the ends of a cable are transmitted to the ground in this manner.

In an arch, the magnitude of the horizontal reaction H varies inversely as the rise h. If a cable, therefore, has the same geometry and loading, it will have an identical value of h at mid-span, although h will be a measurement of sag rather than rise. The horizontal inward pull on the posts will be very large if the line is almost straight, and comparatively small if there is an appreciable sag. Figure 6.4 shows how the initial horizontal forces on the posts cause them to bend inwards. Posts small in cross-section will undergo quite noticeable horizontal deflections at the tops, causing the line to sag more in the middle owing to the shortening of its span. This in turn causes the horizontal reactions at its ends, and therefore the actions on the posts, to be smaller in magnitude.

It may appear to follow automatically that the bending moment at the base of the vertically cantilevered post would reduce in the same proportion. This is not quite true, however, since the inward displacement of the tops of the posts causes an eccentricity in the applied vertical force from the washing line. This effect could be alleviated by initially setting the posts to lean backwards, so that their inward movement brings the posts back towards their intended vertical orientation.

This type of structure, since it exhibits considerable visible deflections, does not belong in the realm of space enclosure; neither for that matter does a long span suspension bridge, which deflects considerably under wind loading. Both reflect, however, the essence of those structural forms where tension is the dominant visible mode of behaviour.

The suspension cable

The fundamental characteristic of a cable is that it responds to applied loading by increasing its length and pulling inwards. That is to say it enters a state of tension, and does not set up any resistance to bending. It behaves as though it has no structural depth, and therefore no capacity to generate a resistance moment. This is easy to accept if the cable is very thin, but is at first sight open to question if the cable, as with large span suspension bridges, has a diameter of about a metre or more. The modulus of any section with a depth of one metre is such that, when multiplied by the permissible stress of the material to yield a resistance moment value, considerable bending moments can be accepted.

A one-metre diameter cable, however, would consist not of a homogeneous mass of steel, but of strands each of no more than five millimetres in diameter. Unless these strands are constrained not to slide relative to one another, they will act as individual elements, and will have no collective identity. Without resistance to horizontal shear, the strands cannot possibly behave as a single structural element with compression on one side and tension on the other, and a neutral axis at mid-depth. The individual strands evolve into the form of a single cable by the process of cable spinning, in which continuous lengths of strand are carried back and forth over the supporting towers until a section of sufficient area has been created. The friction which occurs between contiguous sections of strand by virtue of clamping or connections with vertical hangers supporting the deck is minimal, and is usually ignored.

Cables fabricated before the development of steel strand consisted of short straight lengths connected by couplings. This was a common feature of nineteenth century wrought iron suspension bridges, in which the impression of a parabolic curve is conveyed by the mathematical approximation of many short straight lines.

The catenary

The parabola is the funicular form for the uniformly loaded arch or cable, but only when the load is applied as equal force increments along equal horizontal distances. When a cable of uniform cross section without any additional loads is suspended between two points, it will assume the shape not of a parabola, but of a catenary. This word derives from *catena,* the Latin word for chain. Overhead transmission lines, unused washing lines, and a just completed suspension bridge cable, as yet free of its intended burdens, are all catenaries.

A cable, unless one of a group supporting a load such as the roof of a building or a bridge deck, cannot, in its isolated form be considered as a structural element. In its compressive funicular form it has proved to be a valuable form-finding device for identifying the ideal structural form when the architect has wished to express structure in its purest sense. The Spanish architect Antonio Gaudi (1852–1927) made use of this principle in seeking the most appropriate form for the church in the Güell Colony on the outskirts of Barcelona. The church is in brickwork, so it was vital to find the correct funicular form. By making string models and suspending lead weights to simulate the gravitational dead loads, Gaudi was able to photograph a variety of tensile forms, and use the reversed compressive forms as the basis of his explorations of space. The exact geometry which was finally adopted was neither that of a parabola nor of a catenary, but of a curve somewhere between the two. It was simply the funicular form for the loads represented by the disposition of lead weights in the equivalent tensile model.

The catenary looks similar to a tensile parabola, so the mathematical variation between the curves makes little difference as far as architecture is concerned. The catenary is not one of the conic sections described in Chapter 1, even though it is a function of the hyperbolic cosine, with its mathematical notation 'cosh'. The equation relating the vertical ordinate of a catenary to the horizontal distance x from the origin is

$$y = c.\cosh \frac{x}{c}$$

where c is a non-zero constant.

Two fallacies

The flat cable fallacy

The simplest cable form, depicted at the beginning of this chapter in Figure 6.2(a), in which a load W is supported at the middle, is shown again in Figure 6.5, with added information defining the sag of the cable. The angle between the horizontal axis and the cable can be given the symbol θ, so that the vector T is inclined at θ degrees to the vector H.

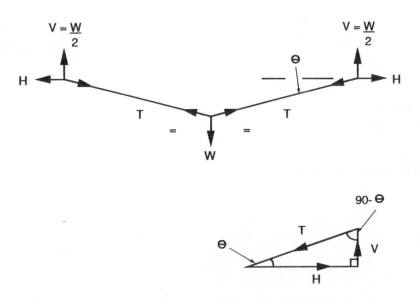

Figure 6.5   *Geometry of point-loaded cable*

Since the loading on the cable is symmetrical, the vertical reactions V at each end will be $W/2$. The triangle of forces for the right-hand support is also shown in Figure 6.5. From the definition of sine given in Chapter 1, T can be expressed in terms of W.

$$\sin \theta = \frac{\text{opposite}}{\text{hypoteneuse}} = \frac{W}{2} \div T = \frac{W}{2T}$$

$$\therefore 2.T.\sin \theta = W$$

$$T = \frac{W}{2.\sin \theta}$$

$$\therefore \text{As } \theta \to 0, \sin \theta \to 0, 2 \sin \theta \to 0, \text{ and } T \to \infty$$

That is, as the alignment of the cable approaches the horizontal, the tensile force T in the cable approaches infinity. Or to be more precise, in order to support any finite load W at its mid-point a horizontal cable needs to be capable of sustaining an infinite tensile force.

From the same force triangle, and from the definition of cosine given in Chapter 1, the horizontal reaction H can be expressed in terms of T:

$$\cos \theta = \frac{\text{adjacent}}{\text{hypoteneuse}} = \frac{H}{T}$$

$$\therefore H = T.\cos \theta$$

$$\therefore \text{As } \theta \to 0, \cos \theta \to 1, \text{ and } H \to T$$

Since T has been shown to approach infinity as the angle θ approaches zero, H will also approach infinity. This means that if a horizontal cable is to support a load W at its mid-point, a horizontal reaction of infinite magnitude will be demanded of the support.

The horizontal cable supporting a central point load is therefore a mathematical impossibility, at least in our world of finite quantities.

The whole of this argument could be repeated for any system of loading, implying that a horizontal cable supporting any load whatsoever, even its own weight, is a pure fiction.

It may be argued that a piece of elastic held with one end in each hand and stretched is both horizontal and supportive of its own weight. This may be true, but this experiment models a different structural form, the prestressed cable. The cable forms discussed in this chapter do not necessarily depend on the initial application of an external tensile force. The vertical and horizontal reactions have been the effects caused by the loading and the cable geometry.

Even a prestressed elastic band, if subjected to a load perpendicular to its axis, will depart from its original straight line geometry. An example of a real life structure erected on this principle is the circus tight-rope. The weight of the tight-rope walker causes a small deflection at every position on the rope at which he balances, but the deflection will be sufficiently large to enable the rope to respond with finite tensile forces. The smaller the deflection, the greater will be the values of T and H, so that great care is needed with the arrangement at the supports. There is nothing alarming about the movement of the rope as the performer crosses from one side to the other, because a different funicular form is generated for each position he occupies. It would be very alarming if the rope remained perfectly horizontal, suggesting an annulment of Newton's Third Law.

The system of vectors in Figure 6.6 is therefore not compatible with the laws of statics, and cannot reflect the loading and support conditions for any real structural form.

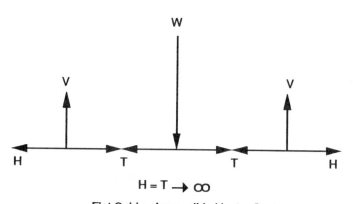

Flat Cable - Impossible Vector System

Figure 6.6   *Fallacy of flat cable*

Although the Newtonian conditions $\Sigma V = 0$ and $\Sigma H = 0$ for the whole structure appear to be satisfied, the horizontal equilibrium state depends on the fallacy of trying to equate two infinite quantities. Moreover, it is impossible to draw a triangle of forces at either of the supports, since there is no force component with which V can be in equilibrium.

The only system which will allow the vertical forces W, V_L and V_R to be in static equilibrium is a beam, where W would be transferred to the support positions by bending and shear instead of by axial tension. The horizontal reactions H would then be superfluous.

The flat arch fallacy

Having shown that there are two theoretical funicular forms for a particular loading condition, one producing pure tension and one producing pure compression, the relationship between the two is of particular interest when the rise of the arch or cable is very small. If the tight-rope were imagined as reflected in a horizontal mirror, the compressive funicular form would appear. The tight-rope walker would, in his mirrored state, be attempting to function as a linear arch walker – a highly dangerous and totally impossible task. Again in a purely abstract sense, the compressive rope form would be unable to maintain its shape, and would immediately revert to the tensile funicular form. Expressed in the laws of mechanics, the rope would pass from a condition of unstable equilibrium into one of stable equilibrium.

The geometry of the notional flat cable, being a straight line, is identical to that of a notional flat arch. Both are dependent on horizontal reactions of infinite magnitude. It is irrelevant whether these reactions are outward-pulling cable reactions or inward-pushing arch reactions. Either way, the structural form is imaginary.

The horizontal reactions for cables and arches can, for a particular span and loading condition, be represented graphically, as in Figure 6.7. The vertical axis is used to plot values of *H*, positive for the arch and negative for the cable, against values of *h*, again positive and negative respectively. The curves are asymptotes (see Chapter 1), being graphs of

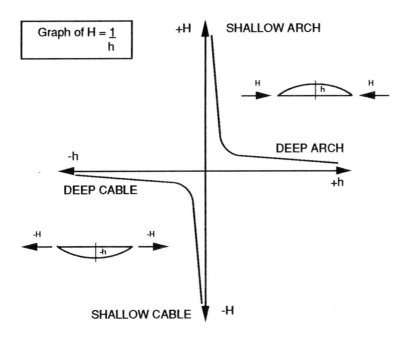

Figure 6.7   *Graph of H = 1/h*

the function $1/x$. They reflect the inverse variation of the horizontal reaction with the rise or sag, the latter expressed mathematically as a negative rise.

H never quite reaches infinity and h never quite reaches zero. At the other limits of the curves, the horizontal reaction H will never quite reach zero however great the sag of the cable or the rise of the arch h.

Cable stayed structures

Principles

One of the most striking introductions into the architectural vocabulary of the later part of the twentieth century is the range of buildings collectively referred to as 'cable stayed'. This description is usually applied to a system which employs steel tensile members, inclined at about 30–45° to the horizontal, above the level of the supported structural element. The object of such an arrangement is to achieve a reduction in the span of this element, and consequently a reduced depth. Figure 6.8 depicts a cable-stayed system in its simplest form, in which the span L of a uniformly loaded beam is reduced to $L/2$ by the introduction of two cables at mid-span. The cables are set at 45° to the horizontal, pass over the vertically extended columns, and down at the same angle to meet the extended beams. The downward-pointing vector arrows at each end are the external forces necessary for vertical equilibrium of the system.

Assuming each of the two $L/2$ metre beam spans to be simply supported, and loaded with a UDL of w kN/metre run, the vertical force to be carried by the cables at mid-span will be

$$2 \, (\tfrac{1}{2} \times w \times L/2) \text{ kN} = wL/2 \text{ kN,}$$

which can be more conveniently denoted by P. By drawing the force triangle for this point, as shown in Figure 6.8(a), the tensile force T in each of the cables can be evaluated, i.e.

$$T = P/2.\sin 45° = P/1.414 = 0.707P \text{ kN}$$

Using this result, the force triangle for the top of one of the extended columns can be drawn, as in Figure 6.8(b). This will have an identical geometry to that in 6.8(a), with the vector arrows pointing in the opposite direction, so that

$$R \text{ (vertical column reaction)} = P \text{ kN,} \quad \text{and}$$
$$T_O \text{ (tension in outer cable)} = 0.707P \text{ kN} \quad (=T)$$

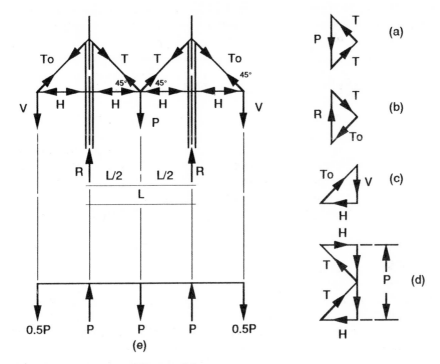

Figure 6.8 *Simple cable-stayed form*

A third force triangle can be drawn at the lower end of the outer cable [Figure 6.8(c)]. Resolving T_O into its vertical and horizontal components,

Vertical reaction at bottom of cable $(V) = T_O \sin 45°$
$$= (0.707P \times 0.707) = 0.5P \text{ kN}$$
Horizontal reaction on extended beam $(H) = T_O \cos 45°$
$$= (0.707P \times 0.707) = 0.5P \text{ kN}$$

The orientation of the vector arrows in Figure 6.8(c) indicates that the direction of the vertical reaction is downwards. This means that a counterweight, or tailing down force, is needed to prevent the structure from rotating towards its centre. The force in the extended beam is compressive. At the beam to column intersection, this force is balanced by a compressive force acting along the whole length of the beam spanning L metres, and eventually by an identical force in the extended beam on the opposite side. The force triangle in 6.8(a) could be redrawn as a force polygon to include these forces [Figure 6.8(d)]. Since these two forces H are equal and opposite at this point, however, the magnitude of T will not be affected.

The line diagram in Figure 6.8(e) indicates the set of vertical forces, generated by and including the original mid-span load P, at the ends of each length of cable. Adding the vectors,

Total downward force = $P + 0.5P + 0.5P = 2P$ kN
Total upward force   = $P + P$ $= 2P$ kN

∴ $\Sigma V = 0$, confirming the system to be in vertical equilibrium.

The effect of taking the cables outside the span L on which the load P was imposed is therefore to produce additional loads at the extremities of the system. These counterweights, each equal to $P/2$, also have to be supported and transmitted to the ground via the two columns. Hence the sum of the vertical reactions at the columns is twice the magnitude of the original load.

Origins and purpose

Structural forms adopted for bridge building are generally determined by purely engineering considerations such as load and span. Bridges are the outcome of responses to the need to cross a valley, road, river or some other interruption to the smooth topography of a landform. Leaving aside structures such as the Poulteney Bridge in Bath, or the Ponte Vecchio in Florence, bridges are not space-enclosing structures. Yet the problem of span is common to both architecture and civil engineering. From the point of view of an informed observer, a bridge is nearly always a clear statement of structural forms both chosen and rejected, in that the support positions are clearly visible. This is not always the case with a building, where the internal system of vertical supports may not be apparent from the outside.

A bridge formed with open webbed steel trusses implies a span too large for a solid beam. An arched bridge, unless chosen for reasons of compatibility or contrast with the surrounding landform or adjacent structures, implies that a trussed system would have been excessively deep. A suspension bridge suggests a span beyond the capacity of an arch. Both of these forms, as described in Chapter 5 and the earlier sections of this chapter, bring into play an enhanced structural depth perceptible only by considering the line of action of the horizontal reactions.

The cable-stayed system also increases the structural depth beyond that of the individual elements of a system. In bridge design, it has tended to provide economic solutions to those bridges just beyond the spans for which an arch would be feasible, and where it would not be considered economic to bring on to site the cable spinning equipment necessary for a continuously curved suspension cable.

What a cable-stayed system effectively does for a bridge is to acknowledge the need for support in those positions where it cannot be provided beneath the bridge deck, such as in a river. Instead, vertical reaction is provided by tensile members from above rather than by compressive members from below.

It is the cable-stayed form rather than that of the parabolic suspension cable that has been infused from civil engineering into architecture since designers came to regard the tensile form as a legitimate response to briefs demanding large spans. One possible reason is the greater diversity of form inherent in the cable-stayed principle. For example, two structures built in the United Kingdom in the last decade, that is since 1980, are arranged on a grid system, each array of columns, beams and cables forming primary structural elements on to which the secondary elements span at right angles. Yet both buildings have their own identity arising out of a unique geometry. The Inmos Factory in Cwmbram, Wales (1982), architects Ahrends, Burton & Koralek, structural engineers Anthony Hunt & Associates, has a cable system which concentrates most of the load on to the central bay. The Sainsbury supermarket in Canterbury (1984), has cables which relieve the interior of the building of the need for support in a similar manner to the system described in the previous section. The form of this building was also conceived by Anthony Hunt & Associates, although the detailed structural design was carried out by Ernest Green & Partners. In both these buildings, additional bays could be added if required, in much the same way that the nave of a Gothic cathedral could be extended westwards. The proximity of Canterbury Cathedral to the supermarket suggests this sympathy between the structural bay arrangement in two buildings separated by some eight hundred years.

Hall Seven of The National Exhibition Centre in Birmingham (1980), (Plate 5), architect Edward Mills, structural engineers Ove Arup and Partners, is, by contrast, more in keeping with a formal classical design in that any addition or removal is not possible without invalidating the whole design concept. Here, the clear 90-metre span is achieved by orienting the cables towards the interior of the building not from just two sides, but from all four. Nicholas Grimshaw's Ice Rink in Oxford (Plate 20), completed in 1984, and for which Ove Arup were also the structural engineers, has another totally different concept in that a single cable system along the central axis of the building is used. This structure is considered in more detail in Chapter 7.

The difference between cable-stayed and suspension systems

A suspension cable carrying a uniformly distributed load is curved in the shape of a parabola. If the loads are applied through closely spaced hangers, the cable follows a set of short straight lines which, as they become shorter and more numerous, approach the parabolic form. A long span suspension bridge, such as that over the Humber with a span of 1410 metres, conveys the impression of a parabolic curve. A closer inspection reveals that the lengths of cable between the hangers are straight. The eye and brain make the transition towards the limiting curve, in which, mathematically, there is an infinite number of straight lines of length approaching zero.

What is common to all these structures is that they are loaded at right angles to their horizontal projection between their points of support. A cable-stayed structure, however, carries no load between its ends except for its relatively insignificant own weight. The cables carry a constant tensile force from one end of the cable to the other, in the same way as a tie in a lattice girder. The suspension cable is also in pure tension, but here the magnitude of the tensile force vector changes as the slope of the cable changes. This force is greatest at the supports, where the slope is steepest, and reaches its minimum value when the slope is zero, that is when the cable is horizontal.

Effects on detailing

The cable-stayed form is identified by its angular elevational geometry. Yet its generic name can be misleading in that the tensile members do not have to be cables made up of thin strands. It is quite in order to use steel of any sectional shape. It could be argued that the main determinant of section shape should be dictated by the ease with which the ties can be connected at their ends. In the National Exhibition Centre structure (Plate 5), the tensile members are rectangular hollow sections, with plates welded to their ends and bolted through to the top flanges of the supported trusses.

Another time-consuming problem with roofs supported from above arises from the need to penetrate the roof at the junction between cable and beam. If the detail is in any way imperfect, the prevention of ingress of water can be a recurring maintenance problem. In an ideal world, a perfectly designed and executed detail can be repeated many times on the roof, resulting in a large number of suspension points and smaller horizontal spans between them. An acknowledgement that faults may occur, on the other hand, could lead the designer to an arrangement with the least possible cables, and therefore the least number of penetrations.

On the outside of a building structured as in Figure 6.8, the tailing down force V must finally rely on a connection to a reinforced concrete element at ground level, whether this be a basement, a group of tension piles, or a calculated mass. A strong visual feature can be made of the transition between steel and concrete above ground level, as shown in Plate 5. Whether such a connection arises in a large or a small structure, taking the steel into the ground would invite corrosion, unless it were housed in an anchorage chamber.

Beam design in cable-stayed systems

The horizontal beam spanning $L/2$ metres is subjected to stresses which originate from two sources. Again assuming a simply supported mode, the maximum compressive and tensile stresses due to bending will be at the mid-span position, that is at $L/4$ metres from its junction with the cable and the column, where the bending stresses will be zero. In addition, the compressive force of $T.\cos\theta$, if applied at the neutral axis of the beam, will cause a uniform compressive stress throughout its length. It is possible to reduce, or even eliminate, the mid-span tensile bending stresses by connecting the cable to the beam so that the axial force is applied in the tensile zone of the beam, in this case below the neutral axis. If this connection is made at a distance of one-third of the depth from the underside of a homogenous rectangular beam, or at the equivalant distance for a composite or non-rectangular section, there will be no tension in the top fibres at either end. The stress condition at mid-span is more complex, since bending stresses are caused not only by the eccentricity of the horizontal component of the cable force, but by the bending moment from the uniformly distributed load w kN/metre over the span of $L/2$ metres. The cable therefore has this secondary function of prestressing the beam, and several trial designs may be necessary to find the most effective position for its connection to it.

If the beam is of reinforced concrete construction, the position will almost certainly be chosen with the object of keeping tensile stresses as low as possible. If a steel 'I' section beam or a lattice truss is used, the critical design objective could well be a minimal compressive stress to reduce the buckling tendency. This would involve connecting the cable to the beam nearer to the underside than the lower middle third position, creating initial tensile stresses on the top surface in opposition to the compressive stresses caused by bending.

Multi-cable systems

Further cables can be introduced if the span needs to be divided into supported lengths of less than one half of the original. Each pair of cables over the roof of the National Exhibition Centre (Plate 5), for example, is so arranged that it divides the roof span L into three effective spans of $L/3$ rather than two of $L/2$.

To reduce effective spans even further, several cables can all meet the column at the top as in Figure 6.9(a). This system is known as the 'fan' configuration, where considerable detailing problems can be caused by the large number of cables meeting at one point. Although all the cables, if they are equally spaced along the beam, will carry the same vertical load, their tensile forces will vary according to their slope. The longer cables, meeting the beam near to mid-span, will have much higher tensile forces than those nearer to the column, not because of their length, but because of the increase in the cosine of the angle of slope.

The configuration in Figure 6.9(b) is the 'harp' arrangement, in which the cables are all parallel to each other, and equally spaced. Since only one cable meets the column in any one position, the detailing is simpler than with the 'fan' system. The tensile force in each cable will be the same, again assuming the beam lengths to be simply supported and the loading to be uniform.

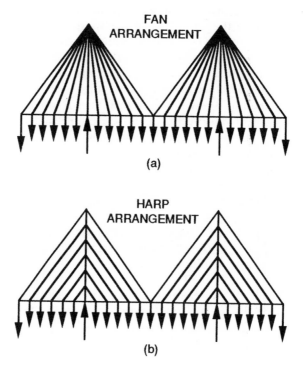

Figure 6.9 *Multi-cable systems*

Both of these arrangements evolve where the beam loading is especially heavy, and excessive beam depths can only be avoided by keeping the intermediate spans as small as possible. The 'fan' and 'harp' configurations are therefore more likely to arise in the design of bridges rather than of long span lightweight roofs.

Lateral stability

When the line diagram of a cable stayed system is completed by showing the columns and the external tailing down cables continued to ground level, the entire form appears to occupy the same structural space as a portal frame. This is especially so with the arrangements in Figure 6.10, in which the two span reducing cables meet the beam to suggest a central hinge. The horizontal distance between each column and its associated external tie in 6.10(a) also suggests the structural depth of a column in a fixed base portal frame. The comparison breaks down when the vector arrows for the horizontal reactions are inserted at the bases. There are no members between the ground and the roof beam capable of providing a horizontal component, and therefore a load path, for *H*. In a steel or timber-latticed column, diagonal members would exist within this space, forming the web and thus creating a composite structural member capable of resisting bending. Unless the column section is made deep enough to resist the bending moment *Hh* at its junction with the beam, thus partly converting the form into a portal frame, additional diagonals need to be provided. These are indicated by dotted lines in Figure 6.10(a), the two tensile diagonals shown with vector arrows in response to the direction of the horizontal force *W*. If *W* were applied from the other direction, the alternate pair of diagonals would act as ties.

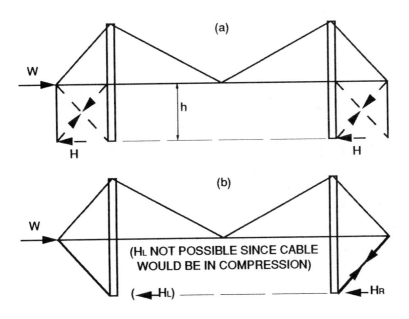

Figure 6.10 *Horizontal forces in cable systems*

In Figure 6.10(b), the external ties are brought down to ground level to meet the column bases and thus acquire a horizontal component and a load path for *H*. The column zone reflects the shape of the bending moment diagram for a three-hinged portal frame. The system may well be stable under the action of purely vertical loads, as the outer surface of the frame will always be in tension, where the ties occur. But the horizontal load *W* shown here will only cause tension in those tie members to the right of the mid-span position. By referring to the bending moment diagram for this load condition in Figure 5.3, it can be deduced that the tie members to the left of mid-span will try to go into compression. An isolated horizontal force in a three-hinged portal frame will always produce tension on the outside of one column and on the top surface of its adjacent rigidly connected beam. The corresponding surfaces on the other side of the structure will be in compression.

The only conditions under which all the cables in Figure 6.10(b) would stay permanently in tension would be when the unwanted compressive stresses on one side arising from horizontal forces were exceeded by tensile stresses arising from vertical dead and live loads. A lightweight roof implies a relatively small dead load. When no superimposed load is present, that is in the absence of snow or any other loads arising from maintenance work, the bending moments on the system and therefore the cable forces would also be of a small magnitude. If high wind forces were imposed on the structure, the cable stress on one side could possibly acquire a net compressive load. This would result in instability not only of the affected cables, but also of the entire structure.

The provision of additional cables to secure resistance to horizontal forces is a common feature in cable-stayed buildings. There are exceptions, such as in the National Exhibition Centre in Birmingham (Plate 5), where the rigidity of beam to column connections can partly or wholly compensate for this deficiency. Where a cable-stayed system is used in conjunction with smaller span buildings, the additional cables provided for resistance to horizontal wind forces and wind uplift are sometimes difficult to distinguish from the main tensile cables in the system. Such is the case with Richard Rogers's Fleetguard Terminal at Quimper in North-Western France (1980), where the cable system reduced the interior clear span of 18 metres to six metres between beam to cable junctions.

Chapter 7

CANTILEVERED AND CONTINUOUS BEAMS

Any architectural feature which is not supported at its extremity can be described as a cantilever. An oriel window, a corbelled parapet and a jettied floor of a medieval timber building all display the principle of the cantilevered form. A more obvious form of cantilever is that which defines the appearance of long span bridges such as the Forth Rail Bridge in Scotland (Plate 21), where the same principle allowed considerable loads to be concentrated at only a few points of support. The 'tree' structure forms discussed in Chapter 5, and indeed trees themselves, consist of horizontally cantilevered branches rigidly connected to vertically cantilevered trunks. A building may evolve with external cantilevers in response to spatial necessity. The way in which the cantilevers are proportioned will invariably have a strong influence on the elevational statements made by the building. For this reason, it is essential that the architect is aware of the constraints imposed on the other structural elements by the adoption of any form of cantilever, and of the possible shapes that the cantilevered elements can be permitted to take.

Cantilevered beams

Bending moments in cantilevers

A beam supported at one end and not at the other is in a condition known as cantilevered. This state usually manifests itself in structural engineering as a projection from a beam which would otherwise be simply supported, or continuous over more than two supports. Cantilevers can occur at both ends of a beam, or at one end only. The structural principles involved in cantilevered forms can be understood within the framework of the calculations for bending moments in simply supported beams. In Figure 7.1(a) the beam, spanning L metres between supports, has two overhanging ends and supports a uniformly distributed load of w kN/metre run. The length of each cantilevered portion is x metres.

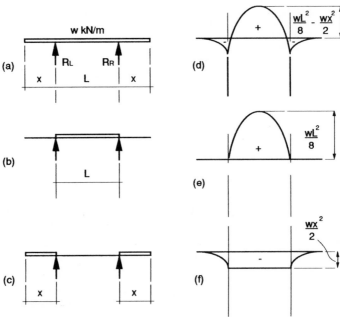

Figure 7.1   *Symmetrical cantilevers – analysis*

If the load is split into two distinct parts, one for the central span and one for the cantilevers, the bending moment diagrams can be drawn separately, and then combined. The separated loads are shown in Figures 7.1(b) and 7.1(c) respectively with the corresponding bending moment diagrams in 7.1(e) and 7.1(f). The combined bending moment diagram appears in Figure 7.1(d).

The bending moment diagram in Figure 7.1(e) is exactly the same as that for a simply supported beam with a span of L metres. Although the cantilevered ends are there, they are theoretically unloaded and do not in this case attract any bending moments. In 7.1(f) the bending moment at each support from the loading on the cantilevers will have a value of $wx^2/2$ kN-m. This follows from the definition of both 'moment' and 'bending moment' given in Chapter 1. The load on each cantilever is

wx kN and the centre of gravity of the load is $x/2$ metres from the support. The bending moment at each support is therefore

$$wx \text{ kN} \times \frac{x \text{ metres}}{2}$$

$$= wx. \frac{x}{2} \text{ kN-m} \quad = \quad \frac{wx^2}{2} \text{ kN-m}$$

The bending moments at positions of a quarter, a half and three-quarters of the length of each cantilever from each end, that is at $x/4$, $x/2$ and $3x/4$ can be similarly calculated as follows:

$$BM \text{ at } \frac{x}{4} = wx. \frac{x}{4} \frac{}{8} \text{ kN-m} = \frac{wx^2}{32} \text{ kN-m}$$

$$BM \text{ at } \frac{x}{2} = wx. \frac{x}{2} \frac{}{4} \text{ kN-m} = \frac{wx^2}{8} \text{ kN-m}$$

$$BM \text{ at } \frac{3x}{4} = 3wx. \frac{3x}{4} \frac{}{8} \text{ kN-m} = \frac{9wx^2}{32} \text{ kN-m}$$

At the ends of the beam the bending moments are zero. Between each end and its nearest support, the increase in bending moment is parabolic, since the magnitude of the bending moment involves a squared term. By drawing the curves below the base line, as in Figures 8.1(d) and 8.1(f), the ordinates acquire negative values.

The reason for the adoption of a negative sign for the bending moments on the cantilevers lies in the convention used in simply supported beams in which the moment about a particular point caused by a reaction R is given a positive value, as adopted in Chapter 1. Any moment tending to cause rotation in the opposite sense must therefore be negative. If the bending moment at the right-hand support in Figures 7.1(c) and 7.1(f) were calculated by taking moments of all forces to the left of that section, the forces to be used in the bending moment calculation would be the left-hand reaction (R_L) and the load on one cantilever. By symmetry, each reaction in this condition where only the cantilevers are loaded will equal the load on one of the cantilevers. Thus,

$$R_L = R_R = wx \text{ kN}$$

From Newton's Third Law, $R_L + R_R = wx + wx = 2wx$ kN

$$\therefore BM \text{ at } R_R = R_L.L - wx(L + x/2) \text{ kN-m}$$
$$= wxL - wxL - wx^2/2 \text{ kN-m}$$
$$= -wx^2/2 \text{ kN-m}$$

The result emerges, as predicted, with a negative value, but it is much simpler to assume from the outset that all cantilever bending moments are negative and to use the simpler calculation outlined earlier. Figure 7.1(f) also shows a constant negative bending moment value between the supports under the loading in 7.1(c). This can be verified by taking moments about any point along the span L. For example, the bending moment at mid-span, caused by the load on either cantilever, will be equal to

$$wx. \frac{L}{2} - wx.\left(\frac{L}{2} + \frac{x}{2} \right) \quad \text{kN-m}$$

$$= \frac{wxL}{2} - \frac{wxL}{2} - \frac{wx^2}{2} \quad \text{kN-m}$$

$$= \frac{-wx^2}{2} \quad \text{kN-m}$$

The final bending moment diagram in Figure 7.1(d) can be drawn by superimposing the positive diagram in 7.1(e) upon the central portion of the negative diagram in 7.1(f). 7.1(d) shows positive values above the original base line, or horizontal axis, and negative values below.

Consequences of cantilevered ends

Three important conclusions can be drawn from the results in the previous section.

First, the cantilevered length of any beam, however it is supported, will produce a bending moment of exactly the same magnitude immediately to each side of the support beyond which it projects. Put more simply, a cantilever moment cannot just disappear. It must be resisted by something, in this case by the beam from which it extends.

Second, the change in sign from positive to negative values of bending moment implies a change in the deflected shape of the beam, as shown in Figure 7.2.

The length of the beam within the positive range of bending moment values will sag, in the same way that a simply supported beam will sag between its two reactions. The lengths of beam within the two negative ranges will hog. Sagging results in tension on the underside of the beam and compression on the top. Hogging produces the reverse effect, with the top side in tension and the underside in compression. The points on the beam at which sagging changes to hogging, and bending moments change from positive to negative, are known as points of contraflexure. The bending moment value here is zero, through which value the graph

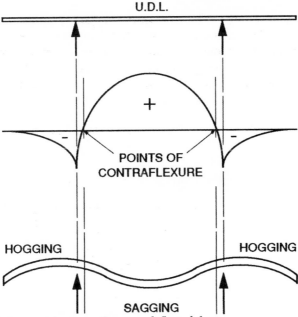

Figure 7.2 *Symmetrical cantilevers – deflected form*

of any function must pass when changing from a positive to a negative quantity. This phenomenon also occurred in the sloping member of the pitched portal frame in Figure 5.7(b).

The third conclusion is that a moment caused by a cantilever brings about a reduction in the positive bending moments within the span. The merging of the bending moment diagrams in Figures 7.1(e) and 7.1(f) has effectively raised the base line of the central parabola by a vertical distance of $wx^2/2$ units. The new height of this parabola, and therefore the maximum positive bending moment in Figure 7.1(d), is

$$\frac{wL^2}{8} - \frac{wx^2}{2} \quad \text{kN-m}$$

At one extreme, if there are no cantilevers, i.e. when $x = 0$, the original midspan bending moment of $wL^2/8$ will be unchanged. The cantilevered beam reverts to the simply supported condition. If, on the other hand, the two reactions are very close together and the cantilevers are long, i.e. x is large and L is small, the mid-span bending moment will have a very small positive value, or even a negative one.

The optimum cantilever

A desired condition when structures of minimum depth are sought occurs when the positive and negative bending moments are equal, or nearly so.

This condition arises when the ratio of the length of the cantilever to the distance between the supports, that is x/L, is approximately one-third, as in Figure 7.3(a). The bending moment diagram in terms of w and L is shown in Figure 7.3(b). The two maximum values, $+wL^2/14.5$ and $-wL^2/18$, can be calculated using the definition of bending moment given in Chapter 1. Thus,

$$R_L + R_R = \frac{5wL}{3} \times \frac{L}{2} = \frac{5wL}{6} \text{ kN}$$

Maximum positive, or sagging, moment at mid-span

$$= +\left(\frac{5wL}{6}.\frac{L}{2} \right) - \left(\frac{w.5L.5L}{6 \quad 12} \right)$$

$$= + \frac{5wL^2}{12} - \frac{25\,wL^2}{72}$$

$$= \frac{wL^2\,(30 - 25)}{72}$$

$$= \frac{5wL^2}{72} = \frac{wL^2}{14.4} \text{ kN-m}$$

Maximum negative, or hogging, moment at supports

$$= -\left(\frac{wL}{3}.\frac{L}{6} \right) = - \frac{wL^2}{18} \text{ kN-m}$$

The algebraic expression will in both cases be the same whether moments of forces to the left or to the right of the section are considered, since both the loading and the geometry of the spans are symmetrical about the centre line of the beam.

There is no question of any of the bending moments on the span L having disappeared without trace as a result of introducing cantilevers. The base line has simply moved up from the springing points of the central parabola, whose maximum height at mid-span remains unchanged at $wL^2/8$ i.e.

$$\frac{wL^2}{14.4} + \frac{wL^2}{18} = \frac{wL^2\,(18 + 14.4)}{18 \times 14.4} = \frac{32.4wL^2}{259.2} = \frac{wL^2}{8}$$

The precise value of x/L at which the maximum cantilever and span moments are exactly equal is 0.352, but the ratio of one-third is

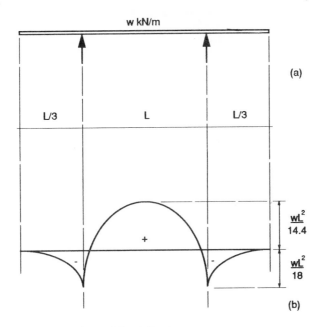

Figure 7.3 *Beam cantilevered by 1/3 of span*

sufficiently accurate for conceptual design purposes. It also has the virtue of being easier to remember.

Referring back to Figure 7.1(c), the loading condition illustrated there is not so abstractive as it may first appear. In a building where the live load is of a varying nature, such as may occur from temporary storage or from sudden crowding, heavily loaded cantilevers together with lightly loaded spans can result in a drastic increase in the deflection of the ends of the cantilevers as compared with the uniformly loaded condition. The bending moment at the support arising from the cantilever load is not affected by what is happening between the supports. However, the absence of the live load between the supports will greatly alter the slope of the beam at the supports, forcing the ends of the cantilevers down. This movement is additional to the purely elastic deflection caused by the cantilever load. In such circumstances the structural engineer may have to advise against this arrangement of supports, that is where x/L equals one-third, particularly where the dead load is light. Such a modification is therefore more likely in structures of timber or light steel framing rather than in those of reinforced concrete. The one-third to one rule is, nevertheless, a sound starting point in architectural design in the quest for a minimal structural depth if supporting columns can be accommodated within an occupied space. Moving columns inwards by even a small distance, giving an x/L ratio of less than one-third, can relieve the architect of the frequently onerous problem of detailing the interfaces between columns and external enclosures. The spaces in between have to be carefully considered in terms of access and circulation by the users of the building. Awkward or unusable spaces should not be created in the pursuit of structural economy.

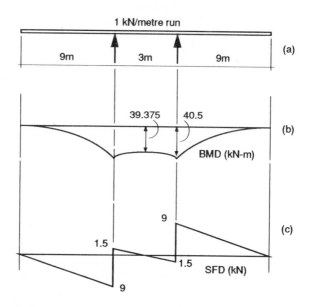

Figure 7.4   *Long cantilevers from short span*

The central corridor

Figure 7.4(a) shows a beam in which the cantilevers are considerably longer than the distance between the supports. The x/L ratio of 3 is significantly high. This form can be generated by an architectural decision to bring all the load on the beam down to two column positions near to the centre of the building. If the column pairs form part of a regular grid, they will form a corridor on the central axis of the building perpendicular to the span of the cantilevered beams. This arrangement, which could be continued vertically through several stories, relieves the exterior of the building of any load bearing functions. There is a penalty to be paid, however, for this freedom from external structure, in the increased depth of the beams, as compared with the provision of an additional column at each end.

The scale of comparison can be most simply expressed by loading the beam with a unit load, that is 1 kN/metre run. Using the general formulae derived in Chapter 1 and earlier in this chapter respectively, the maximum bending moment on each of the 9 metre end spans, if the beam were simply supported at both ends, would be:

$BM = +wL^2/8 = +(1 \times 9^2/8) = +10.125$ kN-m; and if cantilevered,
$BM = -wx^2/2 = -(1 \times 9^2/2) = -40.5$ kN-m

Figure 7.4(b) shows that the critical bending moments for the cantilevered condition occur at the columns. It is interesting to note that in this condition, the bending moment halfway along the cantilever is

$$-(1 \times 4.5^2/2) = -10.125 \text{ kN-m}$$

which is numerically the same as the maximum positive value when the end span is simply supported. The depth of the beam will be governed by the magnitude of the maximum bending moment. Its sign is not relevant where the beam section is symmetrical, but does make a difference in a reinforced concrete section, where the sign will determine whether or not it can function as a 'tee' beam.

The maximum bending moment is four times greater when the end span is cantilevered than when it is simply supported. Since the square of required depth d is proportional to the applied bending moment, i.e.

$$BM = Kd^2 \text{ where } K \text{ is a constant,}$$

d will vary as the square root of BM. A section on which is imposed a bending moment of four times the magnitude of that imposed on another identical section will therefore require an increase in depth of the order of the square root of four, which is two. Put more simply, removal of the end support of a beam whose section is symmetrical about its neutral axis will double the maximum depth required to resist bending. This factor may need to be increased if live load deflections are critical.

The bending moment diagram in Figure 7.4(b) can be completed by first calculating the value at mid-span, that is at 1.5 metres from each support.

$$R_L = R_R = \tfrac{1}{2} \times 1 \times (9 + 3 + 9) = 10.5 \text{ kN}$$
$$\text{Mid-span } BM = (10.5 \times 1.5) - \left(1 \times 10.5 \times \frac{10.5}{2}\right) \text{ kN-m}$$

$$= 15.75 - 55.125 \text{ kN-m}$$

$$= -39.375 \text{ kN-m}$$

The salient values at the ends, the supports and at mid-span, can be joined with smooth parabolic curves. Even at mid-span, therefore, the bending moment is still negative. The entire length of top surface of the beam will be in tension, whilst the entire length of the bottom surface will be in compression. There are no points of contraflexure. The shear force diagram in Figure 7.4(c) indicates that there are three positions of maximum bending moment, and it seems contradictory at first sight that the three bending moment values are all negative. This is perfectly logical when the midspan value of -39.375 kN-m is recognized as the smallest negative quantity and therefore the most positive.

The two 'maximum' bending moment values are in fact numerically almost equal. This suggests that where a long cantilever is projected from a relatively short span, the structural depth required at the support positions must be maintained throughout the distance between them. If this central zone is intended as a corridor, it is not possible to suddenly reduce the depth of the beam to accommodate service runs. Small diameter pipes can be allowed to pass through the beam in the vicinity of the neutral axis, away from the supports. If the beam is of reinforced concrete, holes may sometimes be made above the neutral axis, in the tensile zone, near to the position of zero shear at mid-span. The need to accommodate larger pipes and ducts will inevitably lead to the adoption of a triangulated latticed form, unless a suspended ceiling is provided.

A long span reinforced concrete cantilever will look disproportionately large when such a beam reverts to its unnatural rectangular state from the more natural 'tee' beam condition discussed in Chapter 3. The structural form of such a building could resemble that of a dinosaur, with its head and tail cantilevering from two pairs of closely spaced legs. This analogy may be unavoidable in the case of a bridge. The deep cantilevered form may be regarded as beautiful in long span bridges, where a monumental scale is a natural consequence of its span and loading. The Forth Rail Bridge in Scotland (Plate 21), with its steel tubes forming the lower compression flanges and the lighter sections in tension along the top, is a clear evocation of the laws of structural mechanics. William Morris thought it extremely ugly, and opinions may differ widely in the city of Edinburgh as to its intrinsic worth as a landscape feature. The considerable structural depth of the cantilevers at the supports does not absorb potentially usable space. In a building, a depth of about one and a half metres would be required for a beam cantilevering nine metres. This is more than half the minimum height necessary for a habitable space.

Unless the cantilevers exist as natural, almost organic growths from simply supported beams, with the maximum one-third to one ratio discussed earlier in this chapter, the architect may be well advised to examine alternative spatial arrangements before making a commitment to the long deep cantilever.

The fallacy of the pure cantilever

The foregoing description and analysis of a cantilevered beam has emphasized the identity of a cantilever as an extension of an otherwise simply supported or continuous beam. Cantilevers are frequently depicted in textbooks or engineers' calculations as isolated elements, using the convention of Figure 7.5(a). The hatched lines imply a support which can resist a turning effect, or a moment, as well as a vertical load.

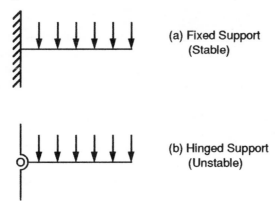

(a) Fixed Support
(Stable)

(b) Hinged Support
(Unstable)

Figure 7.5 *Cantilever notation*

The condition shown in Figure 7.5(b) is clearly unstable, since the single hinged support would immediately rotate in a clockwise direction under the slightest load, even its own weight.

The use of the hatched line notation to depict a fixed rather than a hinged end is a useful shorthand convention when investigating only the cantilevered portion of a structure. The architect should not, however, fall into the habit of assuming that cantilevers can be arbitrarily attached to any structural element. There are certain conditions, apart from extensions to beams, where cantilevers can be introduced without setting the structural engineer an insoluble problem. A reinforced concrete wall, for example, can provide a monolithic connection with a projecting platform or balcony of limited span. The reinforcing bars in the horizontal cantilevered part of the structure can be turned upwards or downwards into the wall, thus ensuring a continuity of resistance moment. A steel bracket can be welded or bolted to a steel stanchion, again ensuring a transmission of bending moment from the horizontal to the vertical member.

What cannot be accepted is the sudden appearance of a horizontal surface projecting from a relatively thin brick wall.

Unlike the continuity achieved where the wall and the cantilever are of the same bending resistant material, a moment connection between two different materials is always problematic. It is possible, with careful design and detailing, to achieve continuity between a reinforced concrete column and a steel cantilever, although this is always easier in the case of a 'tree' structure, in which most of the unbalanced moment arises from

Plate 16   *Nervi's Palace of Labour*   Source: Hassan Falahat

the superimposed loads, as in Nervi's Palace of Labour in Turin (Plate 16), completed in 1961. Since it is the roof which is cantilevered, the discontinuity is in a horizontal rather than a vertical plane. Where one of the elements, such as a brick wall, has little tensile strength, it cannot accept bending moments *per se*. The only possible means of preventing the cantilever from overturning is for the brickwork over the supported end to provide a counterbalance. This would only work if this load from above created the condition in which the resultant vertical force W falls within the middle third of the wall. This eliminates any tension from the face of the wall remote from the cantilever.

This form of construction can only occur in brick walls of considerable thickness, with relatively small projections. This condition is often to be found in the cantilevered treads and landings of staircases in large country houses. An example of this is in Castle Drogo in Devon (Plate 17), designed by Sir Edwin Lutyens (1869–1944) and completed in 1930, in which the span of the cantilevers and the thickness of the supporting walls are both about one metre.

It is far safer to think of a cantilever as a natural extension of a structural form using the same material, rather than as an artificial extension to it.

A curious detail frequently appearing in domestic construction is shown in Plate 18, in which the compression member is noticeably curved. This is illogical when examined in the context of the relationship between bending and resistance moments. Its inappropriateness is even more obvious when it is recalled that the problem with a strut is that any initial lack of straightness is detrimental to its structural role. To start off with a curved member, unloaded between its extremities, seems to be an invitation to buckling. The loads involved here are of a very small order, and this detail, with its vernacular origins, appears to be an acceptable departure from structural logic in favour of decoration.

Plate 17   *Castle Drogo stairs*

Unsymmetrical cantilevers

Propped cantilevers

The beam in Figure 7.6(a), with a fixed support at one end and a simple support at the other, is known as a propped cantilever. It would be difficult to call a cantilever to which a prop has been added by any other name. This form is not statically determinate, in that it has one redundancy. This means that it is possible to release one of the restraints and still leave a structurally stable form. Removal of the prop, or simple support, restores the beam to a cantilevered state as in Figure 7.6(b). Removal of the fixity at the other end leaves a beam on two simple supports as in Figure 7.6(c). This does not mean that a beam designed and built as a propped cantilever can be physically released in this way and still remain stable. As the bending moment diagrams for the three conditions imply, a beam designed for a moment of magnitude $-wL^2/8$ at the left-hand support cannot be expected to suddenly resist a moment of $-wL^2/2$ on removing the prop at the right-hand end. Similarly, if the propped cantilever were suddenly converted to a simply supported beam by releasing the fixity at the left-hand support, its mid-span bending moment would increase to $wL^2/8$ from $wL^2/16$.

Plate 18   *Porch at Otterton*

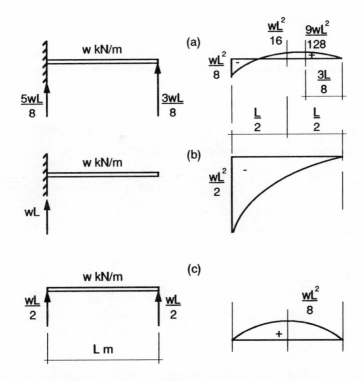

Figure 7.6 *Propped cantilever anaylsis*

The imposing of restraints on the beam additional to those required for static equilibrium means that it cannot be analysed by the laws of statics alone. The results in Figure 7.6 are for a beam of constant cross-section and material. If the beam varied in either respect, the bending moments and reactions would have different values. One method of analysis of a propped cantilever, or any other statically indeterminate form, involves reducing it to a statically determinate condition and examining the influence of the replacement of each redundancy in turn.

The propped cantilever is not so much a deliberately devised structural form as an acceptance of support conditions beyond the control of the designer. A steel beam continuous over several supports and supported on a brick wall at its end cannot behave in any other way. Even a beam forming part of an *in situ* reinforced concrete frame will have end spans closer to the propped cantilever condition than to any other. The end column will, unless very short and squat, support the beam more in the manner of a hinged than a fixed joint, although all monolithic connections will transfer some moment, however small.

Neither can the propped cantilever be said to occur commonly in nature. Organic structures inherent in plant forms tend to cease growing prior to the failure load on their cantilevers being reached. The branches of trees rarely grow so long that they break off at the trunk. Where such a failure is anticipated, artificial supports can be introduced as shown in Plate 19, in which the cantilevered branch is transformed into a propped cantilever, albeit with a further overhang beyond the prop.

Plate 19   *Tree at Killerton House*

Referring again to Figure 7.6, a salient feature of a propped cantilever is the transference of load from the pinned to the fixed end. This outcome is predictable from consideration of the 'pure' cantilever, in which the single support resists the entire vertical load of wL kN (Figure 7.6(b)), with the help of the fixing moment. In Figure 7.6(c), the two pinned ends each resist one half of the beam load. The load distribution between the supports of the propped cantilever would therefore be expected to fall somewhere between these two extremes. As indicated in Figure 7.6(a), five-eighths of the load goes to the fixed end and three-eighths to the pinned end. If the propped cantilever condition were thought of as an anti-clockwise fixing moment applied to a simply supported beam, converting 7.6(c) into 7.6(a), this moment would reduce the right-hand support and increase the left by a force equal to this moment divided by the horizontal distance between the supports, that is by

$$wL^2 \div L = \frac{wL^2}{8L} = \frac{wL}{8} \quad \text{kN}$$

The reactions therefore become

$$R_L = \frac{wL}{2} + \frac{wL}{8} = \frac{5wL}{8} \quad \text{kN}$$

$$R_R = \frac{wL}{2} - \frac{wL}{8} = \frac{3wL}{8} \quad \text{kN}$$

Beams with two equal spans

Consider the beam in Figure 7.7(a) in which a uniformly distributed load of w kN/metre run is applied to two equal spans of L metres, one of which is unsupported at one end. Using the principle of moments to evaluate R_R by taking moments about R_L,

$$R_R. L \qquad = \qquad w.2L.L, \quad \text{from which}$$
$$\text{(anti-clockwise)} \quad \text{(clockwise)}$$

$$R_R \qquad = \qquad 2wL \text{ kN}$$

Since the total load on the beam of $2L$ metres total length is also $2wL$ kN, the result shows that the entire load is carried by the right-hand reaction. This can be verified by taking moments about R_R to evaluate R_L, taking the moment of each loaded span L in turn, i.e.

$$R_L.L + w.L. \frac{L}{2} = \frac{w.L.L}{2}$$
$$\text{(clockwise)} \qquad \text{(anti-clockwise)}$$

$$\therefore R_L.L + wL^2 /2 = wL^2/2$$

Since the term $wL^2/2$ appears on each side of the equation, it can be subtracted from each side leaving

$$R_L.L = 0,$$

and since L is a non-zero quantity, R_L must be zero.

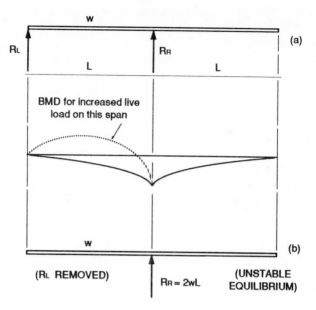

Figure 7.7   *Equal span and cantilever*

The beam therefore appears to be in a state of equilibrium, being perfectly balanced about R_R, rendering the presence of R_L unnecessary. Although this appears logical with R_R being located exactly halfway along the length $2L$ of the beam, this state would be one of unstable equilibrium. Any increase or decrease in the load on either of the spans would alter the reaction values in both of the above equations. An increase in the load on the right-hand, that is the cantilevered, span would bring about an increase in R_R and a negative value for R_L. This would imply that R_L acts downwards to restrain the beam from overturning in a clockwise direction. If the load on the left-hand span, between the supports, were increased, R_R would again increase, and so would R_L which would now have a positive value.

Absence of the left-hand reaction therefore renders the form unstable, as shown in Figure 7.7(b), the form suggesting a child's see-saw rotating about a central fulcrum. Any alteration of the balanced moment condition will cause the beam to diverge from its equilibrium condition. It is a mechanism and not a structure.

Returning to the stable form in Figure 7.7(a), the bending moment diagram for the loading shown is symmetrical about R_R with negative moments throughout the length of the beam. There are no positive bending moments anywhere on the beam. If this arrangement were reflected in the column spacing of a framed structure, the appearance of the beam would suggest the balanced cantilever condition, whether or not the beam were tapered to a smaller depth towards each end. The ratio of cantilever to adjacent span of one is not a sound initial proposal unless site conditions make this unavoidable.

The maximum positive bending moments would occur when the live loads were considered to act between the supports, but not on the cantilever. The identification of positive bending moments, leading to sagging and tension on the underside of the beam, would be particularly important in this form if the beam were to be in reinforced concrete. Design for purely negative moments, with the calculated tensile reinforcement placed along the top surface only, would lead to cracking of the underside of the beam, and possibly failure, in the region of the left-hand support. The dotted line superimposed on the bending moment diagram in Figure 7.7(a) is a possible shape for this graph between the supports when the live load in this region exceeds that on the cantilever.

Continuous beams

Two-span continuous beams

Figure 7.8 shows a beam with the same loading and span lengths as in the previous section, but with a third reaction introduced on the extreme right of the beam. To avoid confusion in the notation, the three reactions in this case have been labelled R_A, R_B and R_C.

With the symmetrical loading and spans as shown, the system can also be perceived as two propped cantilevers, one reflected as a mirror image of the other. Referring to the section on propped cantilevers earlier in this chapter, there will be a maximum negative bending moment of $wL^2/8$ at the central support R_B, which will support five-eighths of the total load, i.e.

$$R_B = 2\,(\tfrac{5}{8} \times wL) = 1\tfrac{1}{4}wL \text{ kN}$$
$$R_A = R_C \qquad\qquad\quad = \tfrac{3}{8}wL \text{ kN}$$

These reactions and bending moment values are dependent on the three supports remaining at the same level. If the foundations on which these reactions are finally supported are built on a subsoil which could allow differential settlement, the distribution of reactions and bending moments could differ considerably. It is therefore vital that the consequences of one reaction moving downwards relative to any of the others are investigated before a statically indeterminate system is adopted.

This is best appreciated if an extreme case is considered. If the reaction R_C settles to an extent greater than the cantilever deflection in Figure 7.7, the beam will have reverted to that condition: R_C will effectively have disappeared. The negative bending moment at the central support would have increased fourfold from $wL^2/8$ to $wL^2/2$, totally invalidating the design assumptions on which the beam dimensions had been determined. Most differential settlements do not involve such a drastic change of identity in a structural element. If such settlement had been foreseen, the design of the foundations would it is hoped have been conditioned thereby, probably resulting in the rejection of isolated pad foundations in favour of an alternative such as a raft or piles. This extreme comparison does, nevertheless, indicate the nature of the alterations to design criteria which differential settlement may cause.

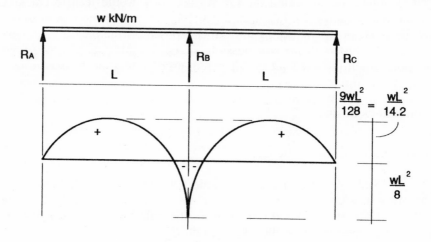

Figure 7.8 *Two span continuous beam*

If the two-span beam is a symmetrical section of steel or laminated timber, it may be sufficient to ensure that the section can resist a bending moment with either a positive or a negative value of $wL^2/8$ anywhere along its length. The beam could then function either as intended (Figure 7.8) or as a pair of simply supported beams sharing a common central support, provided that the change in slope at this position could be accommodated without damaging the integrity of the section. If, on the other hand, there is a possibility that the downward movement of R_A or R_C could be excessive, failure at R_B by the attempted reversion to a cantilevered state could occur before the simply supported mode asserted itself.

By the same token, a two-span continuous beam in reinforced concrete is being asked unreasonably to resist a negative bending moment of $wL^2/8$ with its 'unnatural' rectangular section, whilst the stronger 'tee' section is subjected only to a much smaller bending moment of $wL^2/14.2$. Settlement of the outside supports would further increase this burden on the weakest part of the beam. Fortunately, current codes of practice acknowledge the phenomenon of the redistribution of moments which takes place as the ultimate bending moment value at a particular section is approached. This does allow more of the burden to be borne where the greatest capacity lies, that is nearer to mid-span with the supported slab in compression.

Central spine beams

Continuity at the central support of a two-span beam, with its frequently inconvenient pattern of reactions and bending moments, is usually encountered as an unavoidable constraint on the design of a structure. Cases can arise, however, when the removal of some of the load from the external supports to the centre of a building can be an advantage. An example of this is the roof of Nicholas Grimshaw's Oxford Ice Rink (1984), shown in Plate 20 as an example of a cable-stayed structure. Here, the cable system supports a spine beam on the longer axis of the rectangular building, the four cables breaking up the 72-metre beam length into five equal spans of just over 14 metres. The spine beam extends beyond the enclosed space at each end to equilibrate the vertical and horizontal components of the forces in the cables reaching down from the tops of the columns.

It is in the secondary roof structure, at right angles to the orientation of the cables, that the two-span system appears. The roof members, spaced

Plate 20 *Oxford Ice Rink*

at four-metre centres, are continuous over the central support, that is the spine beam, but connected to the tops of the external columns by hinged joints. These members therefore behave exactly as in Figure 7.8, with an increase in load on the spine beam and a decrease at the hinged external columns. This is to the benefit of this structure because the poor soil conditions led to the decision to adopt piled foundations to the columns and counterweights at each end of the cable-stayed system. The foundations to the columns supporting the hinged ends of the secondary roof beams are pad bases resting on inferior subsoil. The distribution of loads from the superstructure is therefore in keeping with the varying capacity of the two differing foundation systems.

Another building structured on the spine principle, although totally different in appearance, is Eero Saarinen's Yale Skating Rink in New Haven, Connecticut, USA (1959). The spine in this case is an arch, and the secondary structure at right angles to it consists of parallel suspension cables. In contrast to the plane surfaces and nautical appearance of the Oxford Ice Rink, therefore, the form of the Yale arena is that of an upward curving spine supporting downward curving cables. The greater part of the vertical loads are carried by the spine because of the geometry of the cables, the lower ends of which are almost horizontal and therefore have negligible vertical force components. They function as 'half cables' which, in the same way as the 'half arches' discussed in Chapter 5, require horizontal reactions at a different level from that of the main supports.

Deliberate points of contraflexure

Many of the structural forms discussed so far have generated bending moment diagrams containing one or more points of contraflexure. The pitched portal frame in Figure 5.7, the cantilevered beams in Figures 7.1, 7.2, 7.3, and 7.6(a) and the continuous beam in Figure 7.8 all exhibit changes from positive to negative bending moment values at certain positions within the frame or beam. These positions are unique for a particular pattern of loading. They will remain constant if the magnitude of the loads is scaled up or down, provided that a constant factor is applied to all portions of the load. It can be shown, for example, that the point of contraflexure on a propped cantilever carrying a uniformly distributed load occurs at a distance of $L/4$ from the fixed end. If the magnitude of this uniformly distributed load were doubled, the position of the point of contraflexure would remain the same, although the bending moment value at every position would double. If the uniformly distributed load were replaced by a varying load, or a system of point loads, the point of contraflexure would occur elsewhere. It is not always possible to insert a hinge in a structure on the premise that it coincides with a naturally occurring point of contraflexure for an assumed pattern of loading. Even if a uniformly distributed load is taken as the design criterion, it is highly unlikely that the loading will remain uniform at every moment in time. It will probably never be perfectly uniform. A hinge inserted at one of the points of contraflexure in Figure 7.3 would convert the structure to a mechanism, stable only for the unattainable perfectly uniform load.

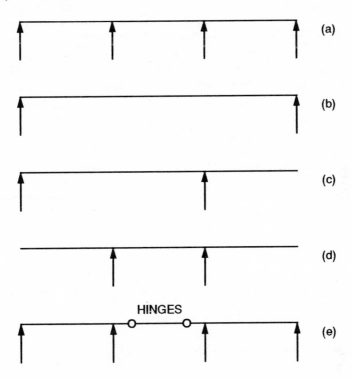

Figure 7.9 *Formation of statically determinate beam*

What is possible within the laws of mechanics is the deliberate choice of hinge positions at the design stage, provided that the occurrence of these hinges does not reduce the structure to an unstable form. Hinges can be introduced to convert a statically indeterminate form to a determinate form, thus making analysis much easier. A statically determinate form can also accommodate temperature changes and differential support settlements without affecting the magnitude of reactions and bending moments.

Figure 7.9(a) shows a three-span beam which, if continuous over both internal supports, would be twice indeterminate. This means that if any two of the four supports were removed, a stable form would still remain, provided that the resistance moment at every section of the beam were compatible with the applied bending moments. This reduction to a statically determinate form could be made as shown in any of Figures 7.9(b), 7.9(c) or 7.9(d). Another reduced form is that in Figure 7.9(e), in which all four supports are retained, but two hinges are introduced into the central span. This removal of the capacity to resist a bending moment by allowing one part of a beam to rotate with respect to the immediately adjacent part is an equally valid way of reducing the statical indeterminacy of the system.

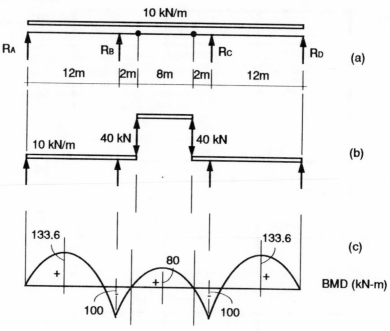

Figure 7.10   *Hinges in central span of three span beam*

It is now possible to analyse the system using the laws of statics. Figure 7.10 shows a beam analogous to that in 7.9(e) carrying a uniformly distributed load of 10 kN/metre run. The system effectively consists of a simply supported central span resting on the ends of two cantilevered beams. The bending moments and reactions for each part of the system can be calculated separately, using the results of the first as part of the loading on the second.

For the central span of 8 metres between the hinges,

$$BM = \frac{wL^2}{8} = \frac{10 \times 8^2}{8} = 80 \text{ kN-m}$$

The reactions on the hinges, which have been left unlabelled to avoid confusion with the four external reactions, will be one half of the load between the hinges:

$$\text{Hinge reaction} = \frac{10 \times 8}{2} = 40 \text{ kN}$$

These reactions will therefore become actions in the form of point loads on the two cantilevers, as in Figure 7.10(b). The maximum negative bending moment on the left hand cantilever supported at R_A and R_B is

$$- (40 \times 2) - (10 \times 2 \times 1) \text{ kN-m}$$
$$= - 80 - 20 = -100 \quad \text{kN-m}$$

Taking moments about R_A,

$$\underset{\text{(anti-clockwise)}}{12.R_B} = \underset{\text{(clockwise)}}{(40 \times 14) + (10 \times 14 \times 7)}$$

$$\therefore 12.R_B = 560 + 980 = 1540$$

$$\therefore R_B = \frac{1540}{12} = 128.33 \text{ kN}$$

Total load on 14-metre part of beam, i.e. 12-metre span plus 2-metre cantilever, is:

$$= 40 + (10 \times 14) = 40 + 140 = 180 \text{ kN}$$
$$\therefore R_A = 180 - R_B$$
$$= (180 - 128.33) = 51.67 \text{ kN}$$

The maximum positive bending moment between R_A and R_B can be evaluated by first locating the position of zero shear, which is at

$$\frac{51.67}{10} = 5.17 \text{ metres from } R_A$$

$$\therefore BM = (51.67 \times 5.17) - \left(10 \times 5.17 \times \frac{5.17}{2} \right)$$

$$= 267.2 - 133.6 = 133.6 \text{ kN-m}$$

The bending moment diagram in Figure 7.10(c) shows the zero values at the predetermined hinges, at which no bending moments can exist under any circumstances. There are also two naturally occurring points of contraflexure on the outer spans. These occur at positions on the beam quite capable of resisting a bending moment if required to do so. Any alteration in the nature of the loading will displace these two points of contraflexure one way or the other.

A practical use of the system shown in Figure 7.10 is in the fabrication of beams off site. If a beam of steel, laminated timber or pre-cast concrete were required for a length of 36 metres, transportation and erection problems may ensue. These problems may well be overcome by fabricating the beam in two lengths of 14 metres and one of 8 metres. They can be connected on site at the hinge positions. Steel hinges would normally be used if the beam sections are of laminated timber or steel. Although steel hinges can be used to connect adjacent lengths of a pre-cast concrete beam, this junction is simpler to construct as a scarfed or halved joint. This is likely to be preferable to fabricating three equal 12-metre lengths, with the problem of accommodating the ends of two beams on each internal column. The location of the hinges is discretionary, and they can be placed much closer to the supports than is shown. The nearer they are to the supports, the greater will be the maximum positive bending moment in the central span, with a consequent reduction in the negative value at the springings of the cantilevers.

Adjacent fully fixed spans

Where a structure consists of a number of equal spans, and the dead load to live load ratio is very high, the behaviour of each span will be very close to that of a beam with two fixed ends. If the negative fixing moments for adjacent spans are equal, their values will change very little when the individual beams are considered as linked to form a continuous beam. Such is the case for the Forth Rail Bridge (Plate 21), in which each support consists of a pair of vertical reactions spaced far enough apart to constitute a 'tree' structure. From these positions, where the structural depth of the trusses is at its greatest to resist the maximum negative bending moment at this position, the cantilevers were built outwards to maintain a state of near balance.

The hinge positions in this bridge occur, naturally enough, at the junctions between the ends of the cantilevers and the suspended spans. The use of relatively short suspended spans as compared with very long cantilevers resulted in very small positive bending moments at mid-span and extremely large negative bending moments at the supports. This is reflected in the enormous structural depth of the bridge at the springings of the cantilevers and the minimal depth of the suspended trusses. The whole system, remembering the analogy between trusses and beams explained in Chapter 4, can be regarded as a continuous beam made statically determinate by the insertion of two hinges near to the centre of each span. The effect of unequal fixing moments from adjacent cantilevers as a result of a train crossing the bridge is to alter the value of each of the vertical support pairs, increasing one and decreasing the other. Because of the magnitude of the permanent dead load of the

Plate 21 *Forth Rail Bridge, Scotland*

structure, both of these reaction pairs always remain positive, that is directed upwards.

At the other extreme in both scale and time is the condition illustrated in Plate 22, part of the ruins of the Roman City of Pompeii in Southern Italy. The entablature resting on the lower four Doric columns consists of seven pieces of stone cut in such a way that the central stone in each span is supported as a wedge on the outer ones. It may appear at first sight that the whole system behaves as a series of flat arches, with the wedging action creating the compressive resistance. This would be plausible if the necessary horizontal reactions at the bases were present. The thrust on the internal columns from adjacent spans is certainly balanced, but there remains an unbalanced thrust on the two external columns. The horizontal reactions to these thrusts could well be provided by the weight of the two outer Ionic columns at the upper level, in much the same way as the pinnacles on the external piers of a Gothic cathedral prevent the arched flying buttresses from moving outwards. There is little possibility of the thrust being resisted internally, since the discontinuity at the joints would inhibit any tying vector near to the soffit of the entablature. The system could, alternatively, behave as a series of three adjacent fully fixed beams, in which the theoretical hinges occur at the sloping joints. The whole structure is statically determinate. The tensile

Plate 22 *Pompeii Ruins*   Source: Derrick Beckett

stresses in the stones arising from both sagging and hogging bending moments would, in view of the small span to depth ratio, be within the limited tensile capacity of the stone. The essential role of all four upper Ionic columns becomes one of creating fixity by preventing the stone below it from rotating about the top of the Doric column supporting it. This total fixity is essential, since the whole length of the entablature would be unstable if it tried to behave as a continuous beam free to rotate about the Doric columns. It may be argued that a Roman colonnade would be more likely to exhibit arch behaviour than that of a trabeated beam system, which one associates more with Greek classical forms. The arch was for the Romans, however, a structural form expressed in a visual rather than a concealed way. It is unlikely that hidden arch action within rectangular elements was part of the Roman vocabulary. Whatever arguments may be advanced as to the precise mode of behaviour, the survival of these structural elements through the destruction of Pompeii by the eruption of Vesuvius until the present day suggests an intuitive sense of structure in the builder which anticipated theories developed much later.

Chapter 8

CIRCULAR AND SQUARE PLAN FORMS

The behaviour of the majority of structural forms encountered in architecture can be understood in relation to the one way spanning systems discussed in Chapters 3–7. The mental habit of analysis is an essential part of an architect's attitude to structure. Analysis by itself may be adequate if the architect is merely appreciating, criticizing or even considering refurbishment of a building. When confronted with the problem of creative design, surely the *raison d'être* of an architect, analysis needs to be accompanied, or even replaced, by synthesis. The creation of a safe structure whose form and proportions do not conflict with the aesthetic, functional and economic ideals of the brief is not a task to be postponed until the later stages of a design. Whether the thought process by which an architect proposes or rejects various possible alternative structural forms is one of deductive or inductive logic, it has to be rooted in a sound theoretical understanding of the underlying principles. This chapter will concentrate on the logical connection between one-way and two-way spanning systems.

Circular plan forms

Curved beams

The problem of roofing a circular space is one that has taxed the ingenuity of architects, engineers and builders from the beginning of civilization. The circle is one of the conic sections described in Chapter 1, in which it was suggested that its effectiveness as a structural element depended on the manner in which it was loaded. A circular beam, as in Figure 8.1(a), needs to be supported at very close intervals if excessive torsional stresses are not to develop. A beam formed from the arc of a circle, as in 8.1(b), will clearly rotate about the axis joining the supports owing to the eccentricity of its vertical load from that axis. Even if it is fixed at its ends by virtue of its continuity around the whole of the circle, torsional moments will arise within the beam, necessitating section sizes greater than those required to resist pure bending.

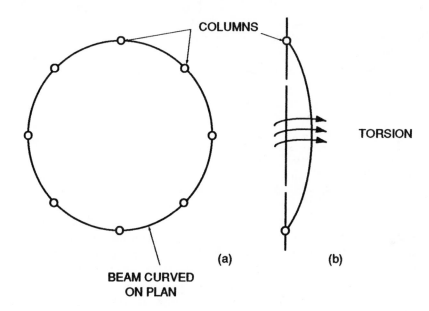

Figure 8.1 *Torsion in curved beam*

A beam strong in bending about the vertical axis is not necessarily capable of resisting torsion. The steel 'I' section, with its opposite flanges in tension and compression, is notoriously susceptible to twisting if subjected to a moment perpendicular to its longitudinal axis. If it has a horizontal curve to start with, these twisting moments are unavoidable. For a section to be able to resist torsion, it needs a similar order of strength and stiffness in bending about both vertical and horizontal axes. The most efficient sections for torsional resistance are square and circular hollow steel sections. If a reinforced concrete section is to be subjected to significant twisting moments, its depth should be of a similar order to its breadth, with additional longitudinal and transverse reinforcement. An *in situ* reinforced concrete slab cast integrally with a curved edge beam does partially compensate for the torsion by virtue of its end fixity. A curved beam on which the supported elements merely rest does not benefit in this way, and relies solely on its sectional properties.

Close examination of structures curved on plan often reveals that the curves do not define the locus of the span. The curved edge is more likely to be the termination of cantilevered elements spanning radially and supported on a polygonal grid set back from the elevation.

Ring tension and ring compression

The circle comes into its own when loaded uniformly in its own plane. Figure 8.2 shows two circles, the loading on each corresponding to an identical in-plane force at every tangential position around the perimeter. In 8.2(a) this uniformly distributed force *t* is applied from the inside and in 8.2(b) from the outside.

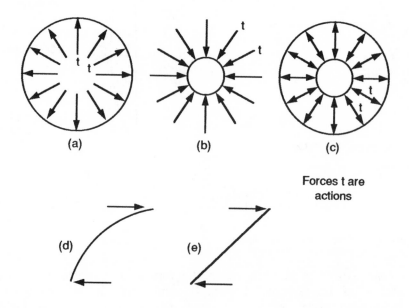

Figure 8.2   *Forces on ring beams*

If the two circles are placed concentrically as in Figure 8.2(c), the forces *t* appear as pairs of vectors caused by the actions of an intervening structural element. Sections through two forms which that element could take are shown in Figures 8.2(d) and 8.2(e), in which the plane of the inner circle is higher than that of the outer circle. The form of 8.2(d), curved also in the vertical plane, is a dome, being the surface traced out by translating the half arch on the circles through 360°. By rotating the straight line in 8.2(e) in a similar manner, the surface of a cone is traced out. If the linear elements in 8.2(d) and 8.2(e) existed in isolation, their structural forms would be the half arch, discussed in Chapter 5, and the inclined beam, which behaves like a ladder resting against a wall. The loads on either of these linear elements would need to be resisted externally by a vertical reaction at the base together with a pair of equal and opposite horizontal reactions, one at each extremity. In the dome and the cone, the forces *t* are resolved into the horizontal circles as ring tension or ring compression. Ring tension occurs in the outer, lower circle in which the set of outward tangential forces *t* causes the circle to expand, thus increasing its perimeter. Conversely, the contraction of the upper, inner circle reduces the perimeter, creating ring compression.

If the dome or cone consists of a series of discrete radial elements, or meridians, the rings will consist of separate structural elements to which their ends are connected. The meridians, if connected only by light cladding or glazing, are free to deflect independently of one another. No circular structural reactions are set up in between the upper and lower rings. For a dome with a continuous surface, either ring tension or ring compression will be present at every horizontal plane, except at one neutral plane where the transition between the two takes place.

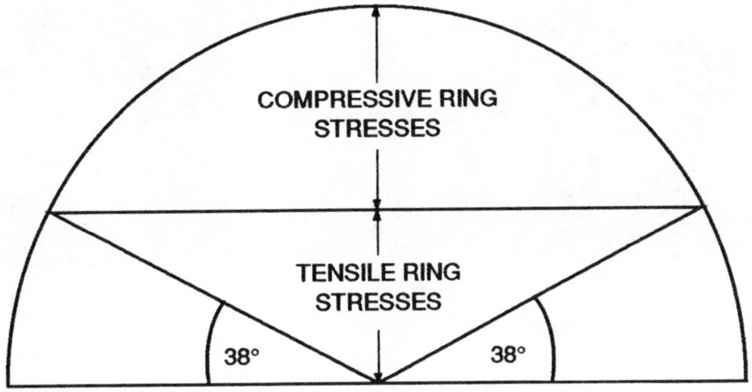

COMPRESSIVE RING STRESSES

TENSILE RING STRESSES

38°   38°

Figure 8.3   *Horizontal stress distribution in domes*

Segmental domes

It can be shown that in a surface dome of constant thickness formed as a hemisphere, the vertical stresses are all compressive, and horizontal ring stresses also occur over the entire surface. As shown in Figure 8.3, the nature of these ring stresses changes at a horizontal plane cut by a line drawn from the centre of the dome at 38° to the horizontal base line. The ring stresses below this line are tensile, and those above are compressive. A reinforced concrete dome which consisted, therefore, of only that surface above this plane would respond to its dead load by compressive resistance in two orthogonal directions.

The advantage of such a dome is that the concrete is being used according to its natural attribute of compressive strength. Bending moments will only arise from asymmetrical loading and would not normally call for any thickening of the dome beyond that required for resistance to the compressive membrane stresses in the curved surface. The majority of the reinforcement is there to resist secondary stresses caused by shrinkage and temperature changes.

Plate 23   *Palazetto dello Sport*

The disadvantage in using only the wholly compressed surface of a segmental dome arises from the paradox that the elimination of ring tension deprives the structure of any resistance to the outward thrust, which will still be present from the 'half arch' behaviour of the meridians. The thrust therefore has to be resisted by external buttressing. Probably the best known example of this system of internal compression and external buttressing is Nervi's Palazzetto dello Sport (1957) in Rome

(Plate 23). The external 'Y' shaped buttresses are integral with vertical columns in the plane of the external enclosure. The columns are therefore visually less prominent than the buttresses, in whose shadow they often appear in illustrations. This conveys the impression that the buttresses form a downward linear extension of the domed surface. In reality the propped 'Y' forms, each with two points of support, are structurally independent and capable of resisting wind forces as well as providing vertical and horizontal reactions to the dome. The ribs of this dome, built using Nervi's 'ferrocemento' process, have a maximum depth of 125 mm for the 58.5 metre span, giving a span to depth ratio of 468 when the surface thickness is considered. Using a similar logic as was applied to trusses and arches in Chapters 4 and 5, the true span to depth ratio, using the rise of the dome from the line of action of the horizontal reaction vectors, is approximately 10.

The dome in architecture

The domed roof evolved as the three-dimensional analogy of the two-dimensional arch in seeking to use masonry so that it remained in compression over most of its surface. The mathematical theories which identify the tensile stresses were not available when the finest masonry domes were constructed. Many of these have been reinforced by iron chains after tensile cracks started to appear. The forms chosen were based on an intuitive feel for the relationship between material and structural form.

Hagia Sophia in Constantinople, 32.6 metres in diameter and finally completed in 537 AD, is a segmental dome which does not define exactly the space beneath. The weight of the circular dome is transferred to the square in which it is inscribed by spherical triangular elements known as pendentives. The plan is elongated on two opposite sides by the provision of two half domes which also resist the horizontal reactions in those directions. There is little provision for horizontal reaction on the other two sides. The splendour of this dome lies in the manner in which light enters through its base, giving the illusion of hovering over its supports. The openings which admit the light occupy the spaces on either side of the ribs, the structural form being closer to that of a ribbed dome than a continuous surface. Horizontal reactions are therefore a necessity, culminating in the addition of a perimeter chain in 1847.

Brunelleschi's Duomo in Florence, completed in 1434, which is octagonal in plan, has the greatest span, about 42 metres, of the domes built during the period of the Italian Renaissance. Marginally smaller is Michaelangelo's dome of St Peter's, completed after his death in 1590. Both have an inner and an outer shell, and both have incorporated a system of chains to act in ring tension. Christopher Wren's 33-metre span dome in St Paul's, England's only Renaissance cathedral, consists of three domes; those visible from the inside and the outside, and a concealed inner surface which is a cone. A vertical section through the lower portion of a cone reveals two straight lines, being the funicular form for a central point load. This is in fact the nature of the loading on the inner dome, whose form is determined by the presence of the masonry lantern, ball and cross at its apex, the weight of which is large in relation to that of the outer timber dome which it supports. The image of the visible outer dome supporting the lantern structure is therefore misleading, in contrast to the dome of St Peter's which has to carry both its own weight and a heavy central point load. Wren also used iron chains at the base of the cone to convert outward thrusts into ring tension.

The alternative to a tensile ring at the base of a dome is to provide external buttresses, which then become part of the architectural statement of the building. In James Gibbs's Radcliffe Camera in Oxford (Plate 24), completed in 1748, the curved buttresses clearly express their intended structural function. Each buttress can be regarded as a vertical cantilever resisting the horizontal and vertical actions from the rib of the dome which it supports. It is dependent on its own weight together with the vertical load from the dome imposed upon it to counteract the tensile stresses arising from outward bending. The eccentric vertical load will produce compression on the inner edge and tension on the curved outer edge. This tendency is reversed by the bending moment caused by the horizontal thrust, which as an isolated force creates tension on the inner face. The buttress is effective if the resultant of the vertical forces remains within the middle third of the section at every level, in which case tensile stresses will be avoided.

Internal planning of circular forms

The prevailing image of the three Renaissance cathedrals referred to in the previous section is undoubtedly the dome in each case. Yet in all three, the domes are surrounded by naves, transepts and porticos which also posed problems of relating structure to space. In none of these churches does the dome exist in isolation. The same applies to Hagia Sophia, even more so as the main dome is dependent on the adjacent half domes for buttressing. In Nervi's Palazetto (Plate 23), the whole of the circular space is enclosed and defined by the dome. The two-dimensional circle has generated the three-dimensional shell structure. Given the sporting function of that building, the circular space is sufficient unto itself. This is not necessarily the case with all buildings whose design originates as a circular plan. In Rome, the two surviving early Christian churches, Santo Stefano Rotondo (built 468 AD) and Santa Costanza (330 AD) have internal sub-divisions based on concentric circles. Gibbs's Radcliffe Camera (Plate 24) is seen by John Summerson (Summerson, 1980, p.42) as derived from Bramante's Tempietto in the cloister of San Pietro, Montorio, Rome (1502), in which two load-bearing perimeters separate the inner dome from the lower flat roof which encircles it. Where a circular building is divided according to the original circular geometry, elegance of style and simplicity of structural form are not difficult to achieve. Structural difficulties are more likely to arise when a circular building is divided radially as though a cake were being sliced, particularly if the slices vary in height to allow light to penetrate the radial sides. The structural complications inherent in such a plan are reflected in uncomfortable and inappropriate internal spaces.

Plate 24   *Radcliffe Camera*

Returning to the present century, two notable reinforced concrete domes, while that form was adopted as the best solution to the large spans, appear from the outside not as domes, but as cylinders. Max Berg's Centenary Hall in Breslau, Poland (1912) expresses its clear 65-metre span internally, but externally shows three telescopic cylinders. The vertical walls of the two inner cylinders are supported on the ribs of the dome. Nervi's 100-metre span Palazzo dello Sport (1959), built after his nearby Palazzetto for the 1960 Olympic Games in Rome, owes its cylindrical appearance to the elevated galleries within the generating circular plan but outside the dome.

Circular plans formed with inclined beams

The extent to which domes, or any other curved shell surface, appear in the architecture of a particular age depends, as well as on the plan forms from which they derive, on factors as diverse as the shape of adjoining buildings and a preference for a particular style. No less important a consideration is cost. As far as generalizations are possible in the economics of construction, curved surfaces are more expensive to form in reinforced concrete than plane ones. The emergence of a cone-shaped space in a material capable of resisting bending moments, using inclined beams, therefore seems inevitable. Anthemius of Tralles, Brunelleschi and Michaelangelo Buonarotti had no other option than to try to discover those forms which would develop only compressive stresses. They were working within the confines of the limited tensile strengths of brick and stone.

The modern architect and engineer have the options of achieving thinner surfaces by using reinforced concrete in optimal surface forms, in which this material will remain substantially in compression, but be capable of some bending resistance when required. They can also use inclined beams arranged in a circle on plan, their upper horizontal reactions mutually transmitted by a compression ring. If the relationship between the inclined beams and the secondary roof structure is such that no radial forces are possible, the cone form will be ribbed, with no interaction between the principal inclined members. Each inclined beam will be subject to gravitational loads over its horizontal projection, causing bending moments of a similar order to those experienced by simply supported beams over the same span. It will also be loaded by horizontal wind forces over the vertical projection. Bending moments so caused have to be considered in conjunction with those arising from vertical forces, but are unlikely to influence the design of the member unless its inclination to the ground is very steep and the roof decking very light.

Given that straight line elements forming a cone will attract bending moments because of their linearity, Wren's choice of a cone for the inner 'dome' of St Paul's seems a curious one given his restriction, late in the seventeenth century, to brick, or unreinforced masonry. But his cone was a surface, not a ribbed structure. Referring to Figure 1.6, a section cut through a cone parallel to one of the sides is a parabola, one of the conic sections discussed in Chapter 1. Figure 8.4 shows two sets of intersecting parabolas, which in Wren's brick cone will produce a purely compressive state when those parabolas are uniformly loaded perpendicular to their bases in their own planes.

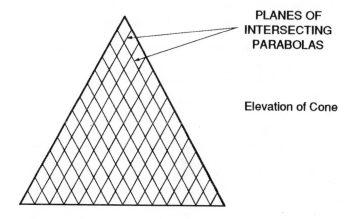

Figure 8.4 *Intersecting parabolas in surface of cone*

The weight of the lantern is carried by the cone in the manner of a radial series of plane trusses, with the inclined members in compression. This produces a prestressing effect which helps to alleviate any tensile bending stresses caused by asymmetrical loading from the weight of the outer dome.

Square plan forms

Direction of span

The domed and membrane systems discussed in the earlier sections of this chapter, whilst possessing a more obvious affinity with the circle, can and have been adapted to the covering of squared and other geometric plan forms. The 206-metre span shell roof of the Centre Nationale des Industries et Technologies in Paris covers a triangular space. On its completion in 1958, this was the largest enclosed span in the world. The 48-metre roof span of Eero Saarinen's Kresge Auditorium at the Massachusetts Institute of Technology (1955) consists of one-eighth of a sphere, that is the curved surface within the intersections of the three great circles. The plan area covered is therefore that of a triangle with curved sides. The pendentives supporting the dome of Hagia Sophia, discussed earlier in this chapter, make the segmental dome compatible with the rectangular plan.

Despite the ingenuity shown in roofing these and many other non-circular spaces, the square and the rectangle are more likely to generate a solution involving linear structural elements. It may be argued that a barrel vault roof constitutes a curved form. This form is, however, a linear structural element which happens to be curved in section, spanning in one direction. Its behaviour is similar to that of the folded plate, in which the tensile and compressive forces of the resistance moment are distanced vertically by the lever arm, but displaced horizontally. This analogy with a simple lattice girder is shown in Figure 8.5.

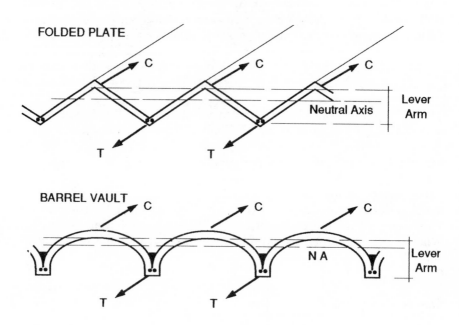

Figure 8.5   *Forces in folded plate and barrel vault*

The instinctive approach to flooring or roofing a rectangular space is to span the shorter distance with the principal structural elements, which are then spaced at centres appropriate to the capacity of the secondary elements spanning between them. A pitched roof on a building with a clear top storey space of 18 metres by 10 metres is more likely, if of traditional construction, to contain trusses spanning 10 metres spaced 2 to 3 metres apart to receive the secondary purlins. To reverse this relationship would result in a built form unusual in appearance, and certainly more expensive. There are notable exceptions where the longer of two rectangular co-ordinates has been chosen as the principal spanning direction, for reasons of aesthetics, internal function or fenestration. One such case is Nervi's Burgo Paper Mill (1965), which resembles a miniature suspension bridge, and another is the 99-metre span arched roof of the Dollan Baths at East Kilbride in Scotland, where the span is aligned with the swimming lanes.

Two-way spanning systems

The deflection of a beam is directly proportional to the cube of its span L and inversely proportional to its stiffness. Stiffness is defined as the product of E and I, where E is the modulus of elasticity, and is a measure of resistance to strain under a given stress, as defined in Chapter 2; and I is the second moment of area, whose magnitude depends on the geometry of the section. In Chapter 4 the z of a section was defined alternatively as the first moment of area. I and z are associated as follows;

z = first moment of area = Ar (determines bending strength of section)
I = second moment of area = Ar^2 (determines stiffness of section)

If the deflection at a point is denoted as Δ and K is a constant whose value depends on magnitude and position of loading, then

$$\Delta = \frac{K.L^3}{EI}$$

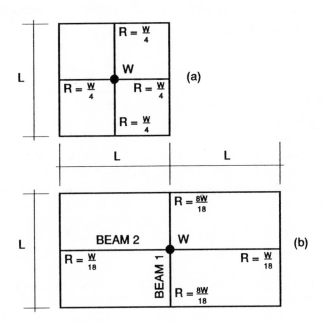

Figure 8.6 *Load distribution in two-way spans*

For the particular case of a point load W at mid-span on a simply supported beam,

$$\Delta = \frac{WL^3}{48EI}$$

If W were placed at the centre of a square bay so that it is supported by two beams at right angles to each other, each of span L as in Figure 8.6(a), the two beams, if prevented from separating at their intersection, would deflect equally.

Both beams have identical E and I values. It can be correctly anticipated, because of the symmetry of the system, that the load will be shared equally between the two beams, one quarter of it reaching each of the four supports. This assumption can be proved correct by writing down the expression for the mid-span deflection of each beam, and making use of the fact of their equality. If the portion of the load W carried by each of the beams 1 and 2 is W_1 and W_2 respectively, then

$$\Delta = \frac{W_1.L^3}{48EI} = \frac{W_2.L_3}{48EI} \qquad \text{where } W_1 + W_2 = W$$

Since all the terms, with the exception of W_1 and W_2 are common to both sides of the equation, cancellation leaves

$$W_1 = W_2 \qquad \text{so that}$$
$$W_1 = W/2 \text{ and } W_2 = W/2$$

Each of the four reactions is therefore $(\frac{1}{2} \times W/2) = W/4$.

If W were supported at the centre of a rectangular bay, as in Figure 8.6(b), with one side twice the length of the other, beams 1 and 2 would have spans of L and $2L$ respectively. This arrangement suggests that the loads W_1 and W_2 will not be equal. Their relative magnitudes can, as with the square bay, be determined by equating the mid-span deflections of each beam, i.e.

$$\Delta = \frac{W_1.L^3}{48EI} = \frac{W_2.(2L)^3}{48EI} \qquad \text{where } W_1 + W_2 = W$$

Cancelling out the denominators, that is multiplying each side by 48EI, and simplifying the term inside the brackets, this becomes

$$W_1.L^3 = W_2.8.L^3$$
$$\therefore \quad W_1 = 8W_2$$

The load carried by the shorter beam is therefore eight times that carried by the longer beam, the whole and partial loads being in the ratio

$$W : W_1 : W_2 = 9 : 8 : 1, \text{ so that}$$

$$W_1 = \frac{8W}{9} \text{ and } W_2 = \frac{W}{9}$$

Applying the formula for the mid-span bending moment under a central point load, derived in Chapter 1, to each of the intersecting beams,

$$BM \text{ under load in Beam 1} = \frac{8W}{9} \times L \times \frac{1}{4} = \frac{8WL}{36} = \frac{WL}{4.5}$$

$$BM \text{ under load in Beam 2} = \frac{W}{9} \times 2L \times \frac{1}{4} = \frac{2WL}{36} = \frac{WL}{18}$$

These results show that even though the beams in the rectangular bay are constrained to deflect equally to render mutual assistance in their supportive task, the shorter beam is doing most of the work. The shorter beam is in fact subjected to a bending moment only 12½% smaller than the $WL/4$ value which it would sustain if supporting the entire load W without any help. The longer beam is contributing very little to the structural system, sustaining only one quarter of the bending moment on the shorter beam.

In the square bay, the bending moment on each of the beams length L is

$$\frac{W}{2} \times L \times \frac{1}{4} = \frac{WL}{8}$$

which is exactly one half of the total value of $WL/4$ on the system.

Grids

The calculation of the bending moments on one pair of intersecting beams involved only one equation. For a grid of closely spaced beams, there will be a much larger number of intersection points, involving one equation for each deflection value. This is a very lengthy process by hand calculation, but relatively simple using appropriate computer software. The principles which emerged from the simple analysis in the previous section are still valid, however many members the grid of beams contains. The fundamental principle is that a grid of members of equal stiffness is at its most effective in creating a load-sharing system when its enclosing sides are approximately of equal length. When the sides are in the ratio of 2:1, a one-way spanning arrangement, with the principal members spanning the shorter distance, and smaller, less stiff members spanning between them, is more efficient. Even when the sides are in the ratio 1.5:1, the critical design bending moments in the shorter members of the grid will still be over twice the value of those in the longer members.

Given that the construction of a grid system in a built form is dependent on each member in the grid being able to pass through all of those members at right angles to it, a two-way spanning system in timber does not seem to be within the art of the possible. Curved forms with intersecting members passing one over the other have been built, but these are the exception rather than the rule. Being domed surfaces, the members would have been more in compression than in bending, and of consequently smaller section.

It may therefore seem ironical that the first plane grid forms used in building were of timber. One piece of timber cannot be made to pass through another without diminishing the section of both pieces, as in a dovetailed joint. What is possible is for a piece of timber to be supported on one side of another, the shear force being transferred by nailing through the breadth of the supporting member into the end grain of the member it supports. This is a common detail in domestic construction used for trimming around fireplaces and stairwells, although the end grain is not the most efficient position on which to impose bearing stresses from nails or screws.

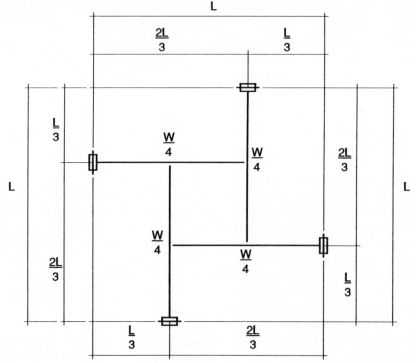

Figure 8.7 *Mutually supported joists*

Figure 8.7 shows an arrangement of four mutually supported joists, in which a total load of *W* is centrally supported over a square bay with a side length *L.* The geometry of the grid is arbitrarily chosen so that each of the four members is loaded and supported at points whose co-ordinates are *L*/3 and 2*L*/3 from its external support respectively. The central square therefore has a side dimension of *L*/3. Because the loading and the geometry are symmetrical, the value of the point load at each of the four intersections, and also of each of the four reactions, is *W*/4. The bending moment at each intersection is therefore

$$\frac{W}{4} \times \frac{L}{3} = \frac{WL}{12}$$

If the enclosed square were structured by four pieces of timber of lengths other than 2*L*/3, then the changing geometries of the system could be imagined to vary as the aperture of a camera. When the aperture size is greatest, that is when the central square has the same dimension *L* as the enclosing square, the point loads *W*/4 coincide with the support positions, and there are no bending moments in the system caused by *W*. If the aperture size becomes very small, the bending moments increase. When the central square disappears as its four corners converge, *W* is no longer divided and becomes a single point load. Each reaction is still *W*/4, so that the maximum bending moment, calculated by taking the moment of any reaction about the centre, is

$$\frac{W}{4} \times \frac{L}{2} = \frac{WL}{8}$$

This condition is exactly the same as that considered in the previous section and illustrated in Figure 8.6(a), the *WL*/4 bending moment for a central point load on a one-way spanning beam being exactly halved.

The latter condition is clearly not attainable with timber for the reasons stated earlier. The use of the larger central square was the basis of grid systems proposed by the Renaissance architect Sebastiano Serlio (1475–1552) (Serlio, 1982, Fol.12) and, in a modified form, of Christopher Wren's mathematics tutor John Wallis. Wren adopted a grid system for the painted ceiling of the Sheldonian Theatre in Oxford, completed in 1668, by using a pattern of short lengths of timber in mutual support until the external walls were reached (Plot, 1672, pp.272).

Reinforced concrete waffle slabs

Of all of the commonly used structural materials, concrete is, in its *in situ* method of construction, the most conducive to the formation of a grid system. At the right-angled intersection of two rectangular ribs there is a square zone, bounded by the breadth *b* of each of the ribs, common to both, as shown in Figure 8.8. When subjected to positive bending moments along the axes of the ribs, the concrete above the neutral axis is compressed biaxially. The orthogonal tensile reinforcement near the bottom of the ribs, which must cross with one layer above the other, ensures continuity of resistance moment in both directions.

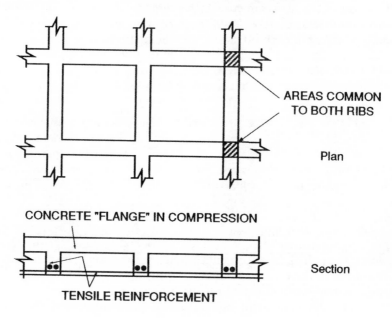

AREAS COMMON TO BOTH RIBS

Plan

CONCRETE "FLANGE" IN COMPRESSION

Section

TENSILE REINFORCEMENT

Figure 8.8   *Reinforced concrete grid*

This principle applies whether the ribs are the members of a two-way spanning deck such as a hollow tile slab, continuously supported along the boundaries by walls or by deeper beams, or whether they are the members of a horizontal structural element of constant depth. The latter form of construction is commonly known as a 'waffle slab'. A waffle is defined in the dictionary as a kind of batter cake, whose form is characterized by a symmetrical arrangement of square voids. In the reinforced concrete equivalent, the voids are created by placing pre-formed moulds on the horizontal formwork, the spaces in between defining the orientation of the ribs.

In a hollow tile slab, square on plan, and assumed to be simply supported on edge beams or walls, the design can be based on a maximum bending moment at mid-span of $wL^2/16$ which is one half of the value for a one-way spanning slab, where each metre width would sustain a bending moment equal to $wL^2/8$. This follows the same logic as the reduction from $WL/4$ to $WL/8$ for the point load discussed in the section before last. In a waffle slab, supported only at the corners of each bay, where all the ribs have identical sections, the maximum bending moments will occur on the lines joining the columns. The ribs in these zones, or 'beam strips', could be made deeper to accommodate the higher bending moments, but this would be to the detriment of the appearance of the soffit of the slab. Even if the slab were concealed by a suspended ceiling, downstands could obstruct the paths of service ducts. If the integrity of the geometry of the form is to be preserved, the variation in the required resistance moments must be conditioned by the amount of tensile reinforcement. Although the ribs may have identical external appearances, therefore, the internal characteristics of each one depend on the demands made upon it. This is a feature common to all elements made of reinforced concrete.

Where the ribs carry negative bending moments, with the tensile reinforcement coinciding with the flange of the 'Tee' section, the compressive force is restricted to the breadth of the rib. This is as unavoidable here as in any continuous system in reinforced concrete. Apart from providing a compressive soffit slab in this zone, the resistance moment of the offending rib has to be enhanced as necessary. In extreme cases, this can be achieved by the application of a post-tensioned prestressing force after the concrete has hardened to its design strength. If this force is applied as an eccentric load, it is possible to maintain the entire section of the rib in a state of compression.

A grid of reinforced concrete ribs with total voids in between, in which there is no enhancement of the resistance moment beyond that of the rectangular section, is familiar on a small scale in the construction of pavement lights, where the intervening spaces are filled by blocks of glass. It is also a possible solution where services need to be passed vertically from one storey to another. In Richard Rogers's Lloyds Building in London, completed in 1986, Ove Arup and Partners, the structural engineers, used this system to link floor and ceiling service zones. Those ribs attracting high bending moments were selectively prestressed, and by so doing the rib depth for the 10-metre maximum span was kept to 500 mm.

One of the most critical features in waffle slab design is the transfer of the load to an isolated supporting column. If the ribs in the beam strip were allowed to continue as rectangular members up to the column face, the column, in providing its reaction to the load, would be in danger of punching through them. The failure mechanism would be that of shear, manifested by a diagonal tension failure in the concrete as described in Chapter 1. For this reason, a solid zone acting as a column head within the depth of the ribs is always necessary. As well as easing the load transfer between slab and column, a higher resistance moment is achieved in an area where the negative bending moments may be difficult to absorb into the isolated ribs. The column head may, if an analogy with Classical forms is sought, be regarded as a twentieth century extension to Alberti's five orders listed in Chapter 3.

Steel space decks

A steel space deck can be envisaged as the equivalent in steel of the reinforced concrete waffle slab. A steel gusset plate can form a nodal connection in a plane truss or lattice girder, resolving force components into two orthogonal directions. A three-dimensional nodal connection can transmit force components into three orthogonal directions, thus making it possible for plane trusses to intersect. At a node point receiving four horizontal members, four diagonal members and one vertical member, there will be nine axial forces, either compressive or tensile, resolving themselves into components to keep the node in equilibrium.

The geometry of a space frame is therefore a series of square based pyramids, the apexes of which are joined by horizontal members in two directions at right angles, as shown in Figure 8.9(a). If the pyramids are arranged in a single line with apexes joined, as in Figure 8.9(b), the resulting configuration is a one-way spanning lattice girder, triangular in section. The tensile force in the resistance moment couple is shared by the two longitudinal bottom members.

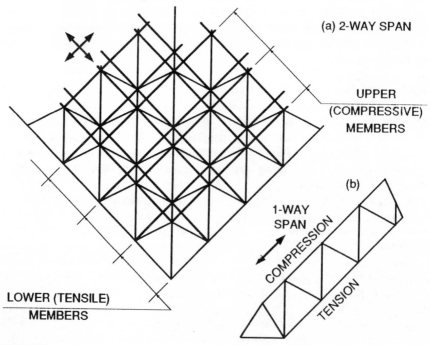

Figure 8.9 *Steel one and two way lattice systems*

Space frames designed and constructed by a particular fabricator can usually be identified by the form of the nodal connector used. Whereas in a waffle slab the structural topping, cast integrally with the ribs to form a 'Tee' beam, also forms the structural decking of the floor or roof, a steel space frame exists as a separate entity. If the top members of the frame are arranged in one metre squares, a dimension typically used, a preformed secondary element must be provided. This may be flat, necessarily so in the case of a floor. In a roof, a pyramidal form may be used to optimize the ingress of natural light. Here, the structural form is expressed indirectly rather than directly, in that the secondary elements, not the primary ones, are seen externally.

Following the logic of the preferred square grid, a rectangular plan form is more conducive to a one-way spanning system, with deep lattice girders spanning the shorter distance. If a two-way spanning solution is preferred in the pursuit of minimal structural depth, a plan area with a two to one side ratio can be converted to two square bays by the introduction of columns half way along the longer sides. Space frames are rarely economical for spans of less than 25 metres. Most systems are based on a module corresponding to about one metre in depth, and a span to depth ratio of about 25 to 30 usually yields a solution in which the individual members are relatively small.

It is not difficult in space frame design to accommodate the higher bending moments in the beam strips on the column grid. There is ample scope for variation in the resistance moment between two sections in a space frame of identical external appearance. Since the resistance moment couple is the product of the force in the horizontal flange and the lever arm, i.e.

$$RM = F \times L.A.$$

and F is proportional to the cross-sectional area, RM will increase in that same proportion. If the tensile and compressive flanges consisted of square hollow sections 100 mm by 100 mm, the thickness of the steel could vary from 4 mm to 10 mm, with section areas of 15.3 cm² and 35.5 cm² respectively. The resistance moment of two sections in which these extreme sizes were used would therefore vary by a ratio of 2.3.

Cantilevers in grid systems

Figure 8.10 shows a square grid in which the four columns have been set inside the enclosing walls, leaving cantilevers on all four sides.

If this has arisen as a result of a decision to separate the columns from the enclosing walls, whilst still maintaining the greatest possible interior free space, the bending moments on the cantilevers will be very small compared with those in the span. This arrangement could apply to a sports hall or any other arena where internal columns must be excluded.

If the frame is to cover a space for which the architect is free to choose that column spacing which will yield the minimum structural depth, the structural designer will seek an equality between positive and negative bending moments. This could apply in a building where the grid lines define internal spaces without the columns acting as obstructions, such as a market hall or exhibition space. Referring to Chapter 7, a cantilever to span ratio x/L of one-third in a one-way spanning element was shown to be the most efficient. This is also a reasonable starting point in the planning stage for a two-way system. This may need to be modified by the structural engineer responsible for the detailed design calculations for two possible reasons.

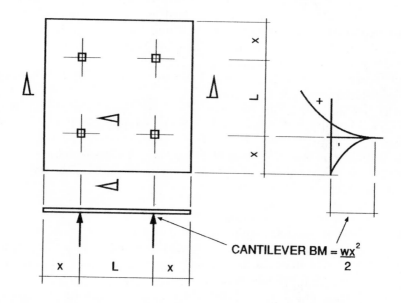

Figure 8.10　*Cantilevered two way span*

First, those areas of the space frame or waffle slab outside the column grid, that is the cantilevers, will mostly tend to span in one direction rather than two. Only those ribs or truss elements in the corners are constrained to span in two directions, their behaviour being of a complex nature in that they are effectively cantilevered from other cantilevers. Most of the other cantilevered elements will receive little assistance from the much longer elements at right angles to them, and will develop negative bending moments approaching the full cantilever value of $wx^2/2$.

For a cantilever to span ratio of one-third, therefore, the structural depth of the grid may have to be increased to accommodate the negative bending moments.

Second, a waffle slab with cantilevers will develop negative bending moments in the ribs, where the resistance moment will be less than that for the 'Tee' section. Means of compensating for this were discussed earlier in this chapter, but long cantilevers from any reinforced concrete members where the 'flange' is placed above the neutral axis are not likely to lead to the most efficient structural forms. In spite of this, there are cases where the aesthetic quality of a coffered ceiling overrides the pursuit of the most efficient structure. In Denys Lasdun's National Theatre in London (1976) the penetration of the first floor waffle slabs into the foyer on the ground floor exploits to the full the creative possibilities of internal cantilevers.

The hyperbolic paraboloid

Geometry

In Chapter 1, this shape was described as being generated by a pair of one of the conic sections, the parabola. Two branches of another conic section, the hyperbola, appear at every horizontal section. On the surface lie two intersecting sets of straight lines, which coincide with those vertical sections where the curvature changes from upward to downward, and vice versa. If the hyperbolic paraboloid, or 'hypar', were square on plan, and the two generating parabolas had the same equation, these straight lines would be orthogonal and at 45° to each parabola. All the straight lines in each set would also be parallel to each other, making the surface constructable with straight lengths of timber. The timber could be either the chosen material for the built form, or the formwork to another material suitable for moulding.

A hypar is therefore admirably suited to timber or to reinforced concrete. The lines of upward curvature will behave in the manner of arches, and those of downward curvature in the manner of suspension cables. If the hypar is of reinforced concrete, the principal tensile bars will be aligned with the downward curves. If it is constructed of straight lengths of timber, at least two layers of boards must be provided so that there are two component directions in which to resolve the resultant tensile forces on the downward curve.

The single hypar

A hypar can be used as a single structural element to cover a square plan by supporting it at its two lowest points, that is where the whole of the outward thrust is concentrated, as shown in Figure 8.11. If these two positions are at ground level, the horizontal reactions H can be provided by a tie member within the floor in the same manner as for a linear arch or portal frame.

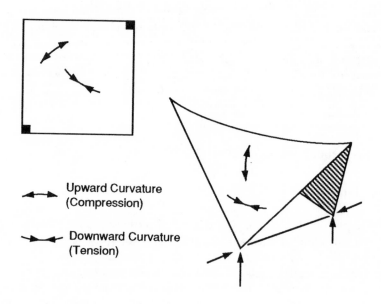

Upward Curvature
(Compression)

Downward Curvature
(Tension)

Figure 8.11   *The hyperbolic paraboloid*

If the bases were pinned, the form would be in unstable equilibrium, since the slightest eccentricity of load about the line joining the bases would cause the structure to rotate about that axis. Stabilization against this tendency, and against wind loading, can be restored in one of two ways.

The first method involves tying the hypar down at each of the other two corners, the highest points. Only one tie would be active at any moment of time, depending on the direction in which the hypar was trying to rotate. The neatness of this solution could depend on how well the ties are concealed by corner window mullions, assuming that the reason for choosing the hypar form was a clear expression of structure through glazed enclosures. The bases could then be physically expressed as true hinges.

The second method is to use fixed instead of hinged bases, the greater dimension being perpendicular to the line joining the bases. This solution contains the paradox of being less elegant than the first, yet more expressive of the structural behaviour of the hypar under all conditions of loading.

Combinations of hypars

A more ordered method of using the hyperbolic paraboloid form is to use them in groups of four rather than in isolation. Figure 8.12(a) shows a square plan with a column at each corner, each of which supports the lowest corner of one hypar. The other three corners of each one are all at the same raised level, forming two horizontal ridges along the axes of the plan. All four elevations display a pedimental form.

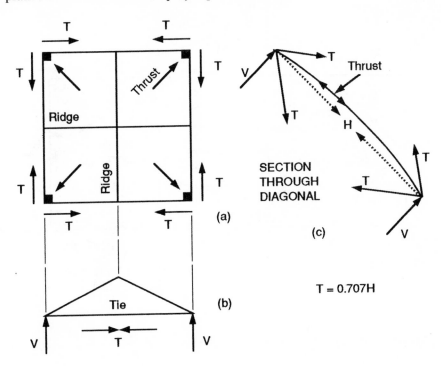

Figure 8.12 *Four hypars with corner columns*

The principal compressive arch curves therefore cross diagonally, meeting at the centre. Unless the columns are designed as vertical cantilevers capable of resisting the diagonal thrust, the tops of the columns need to be tied as shown in Figure 8.12(b) and 8.12(c). The diagonal thrust is resolved horizontally on plan into the components in the two intersecting ties, so that if the required horizontal reaction equals *H* kN, the force *T* in each tie is such that

$$T = H \times 0.707 \text{ kN}$$
$$\text{since } \sin 45° = \cos 45° = 0.707.$$

This arrangement therefore demands no tie members within the enclosed space. Rain water will drain towards the four corners.

An alternative method is to locate the columns at the mid-point of each side, as shown in Figure 8.13(a), allowing the raised corners to be free of any structural elements. The columns again need to be tied to absorb the resultant outward thrusts, which would be perpendicular to each of the four sides of the square, as shown in Figure 8.13(b). In this case the force in each tie will be equal to the thrust from one hypar element.

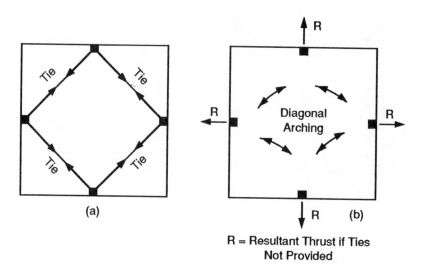

Figure 8.13   *Four hypars with unsupported corners*

The ties in this arrangement occur by necessity within the enclosed space. Drainage is again towards the columns, this time on the orthogonal axes of the plan.

One of the first timber hyperbolic paraboloid roofs in the United Kingdom was erected for the Wilton Royal Carpet Company in 1957 (TRADA, 1984, p.99). The columns were arranged as in Figure 8.13, the dimension of the covered square being 35 metres. The total thickness of the three layers of timber was 45 mm, so that the ratio of span to thickness of material used was 35000/45 = 778. As reasoned in Chapters 5 and 6, however, the true span to depth ratio of a structural system should reflect the rise of an arch or the sag of a cable. For a group of hypars supported as in Figure 8.13, this would be the height of the saddle point, measured from the tops of the columns.

The hypar umbrella

This is the name usually given to a group of four hypars supported on a single central column, as shown in Figure 8.14(a). The concept of true structural depth is important in the understanding of the working of this system.

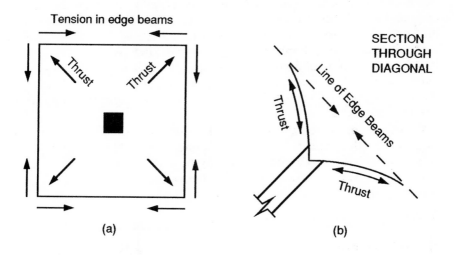

Figure 8.14 *Four hypars with single central column*

By vertically displacing the high and low points of the compressive diagonal of each element, a structural depth is generated. The diagonal section through the structure in Figure 8.14(b) resembles a balanced cantilever, the arch curve taking the role of the compressive force in the resistance moment couple. The outward diagonal thrust at each corner is resolved into inward pulling force components along each pair of orthogonal straight edges. Conversely, by resolving these components back into the diagonal in the same horizontal plane, the resistance moment couple is completed. The outer edges must be made continuous to transmit these tensile components, so that the junction between adjacent hypar elements is particularly important. If the structure is of *in situ* reinforced concrete, this is easily achieved internally by providing adequate bond lengths for the steel bars in the edge beams. If the four elements are of timber, the tensile forces must be transmitted from one edge to the next by an assembly of steel plates and connectors.

This form is just as much a tree structure as a unit of Nervi's Turin Palace of Labour (Plate 16). The essential difference is that in Nervi's plane cantilevers, the structural depth is visible, whereas in the hypar umbrella it is only perceived by conceptual analysis. Also, Nervi's column to cantilever junctions permit the smooth transition of bending moments arising from unbalanced vertical loads. The hypar umbrella only behaves as described above in the unstable equilibrium state, that is when the bending moments on either side are exactly equal. Because the line of the resultant tensile force only passes over the column rather than being physically connected to it, there is no means by which this structural depth can cope with unbalanced moments. Whether these are caused by unequal dead loads, by unevenly distributed live loads or by alternate pressure and suction areas on the roof surface, the unbalanced moments can only reach the column via the connected lower surface. This would therefore need to be of a greater thickness than that required to resist the purely compressive or tensile stresses in the hypar element. Alternatively, the column can be extended above the roof and vertical diaphragms provided around it.

Whichever measures are adopted, the suitability of *in situ* reinforced concrete for this type of monolithic connection makes it preferable to timber for this form. Drainage can only be towards the central column, in or around which a down pipe has to be concealed.

Alternative hypar forms

Although the hyperbolic paraboloid has been examined as a solution to the square plan, it would be misleading to suggest that other configurations are not possible. Individual hypars can be non-rectangular, and these can be combined to cover a diversity of forms including triangles and hexagons. To conclude at the point where this chapter began, one of the most imaginatively conceived circular structures ever built was the Priory of St Mary and St Louis in St Louis, Missouri, USA (Cowan, 1978, p.167), designed by Hellmuth, Obata and Kassabaum. The 43-metre diameter space is structured by three ascending tiers of hypars, none of which has an individual span exceeding 6.4 metres.

Although it could be argued that hypars in a group lead to more rational solutions than when used as single elements, the beauty and simplicity of the form are far more sharply expressed as one unit. The mathematical description of the surface may be that of a saddle or mountain pass, but the elevation of a roof consisting of a single hyperbolic paraboloid is more redolent of a bird in flight. Whatever the current vogue in architectural styles, this imagery will always be seductive for those architects conscious of the expressive power of structure.

BIBLIOGRAPHY

Clifton-Taylor, A., *The Cathedrals of England,* Thames and Hudson, 1967.

Cowan, H., *Science and Building – Structural and Environmental Design in the Nineteenth and Twentieth Centuries,* John Wiley and Sons, 1978.

Feynman, R., *The Character of Physical Law,* MIT Press, 1967.

Plot, R., *The Natural History of Oxfordshire,* Oxford, 1672.

Serlio, S., *The Five Books of Architecture, First Book, First Chapter,* Dover Publications Inc., 1982.

Summerson, J., *The Classical Language of Architecture.* Thames and Hudson, 1980.

Summerson, J., *Georgian London,* Penguin, 1978.

Ruskin, J., *The Stones of Venice,* Faber and Faber, 1981.

Thompson, D.W., *On Growth and Form,* Cambridge University Press, 1961.

TRADA, *Timber in Construction,* Batsford/TRADA, 1985.

Index